Supplementary Exercises

for

Principles of Chemistry

Instructor's Version

Supplementary Exercises

for

Principles of Chemistry

Instructor's Version

MICHAEL MUNOWITZ

W • W • NORTON & COMPANY NEW YORK • LONDON

PRINTED IN THE UNITED STATES OF AMERICA

ISBN 0-393-97550-9

W. W. Norton & Company, Inc., 500 Fifth Avenue, New York, N.Y. 10110
http://www.wwnorton.com

W. W. Norton & Company Ltd., 10 Coptic Street, London WC1A 1 PU

1 2 3 4 5 6 7 8 9 0

Contents

Solutions appear immediately after the Supplementary Exercises for each chapter.

Chapter 1

Fundamental Concepts

S1-1. Light travels at a speed of 3.00×10^8 m s^{-1} in vacuum. Calculate the equivalent value in each of the following sets of units:

(a) kilometers per minute
(b) centimeters per week
(c) miles per hour
(d) inches per year

S1-2. The dimensions of a certain cube are 5.00 cm × 5.00 cm × 5.00 cm. Calculate the volume in liters.

S1-3. Calculate the volume, in milliliters, of a sphere that has a radius of 1.00 m.

S1-4. The *acre*, a unit of area, is equal to 4840 square yards. **(a)** Calculate the equivalent value in square feet. **(b)** Calculate the equivalent value in square meters.

S1-5. Confined at atmospheric pressure and a temperature of 0°C, a gas containing 6.0×10^{23} helium atoms will fill a cubical container 28.2 cm on a side. How many atoms are distributed per milliliter?

S1-6. Assume that a body has a net positive charge of 1.00 C. How many electrons must be added to make the body neutral?

S1-7. Two stationary particles, each with a charge of –2.00 C, interact at a distance of 1.00 cm. **(a)** Calculate the electrostatic force. **(b)** At what distance will the electrostatic force fall to 6.25% of its original value? **(c)** At what distance will the electrostatic force be tripled relative to its original value?

S1-8. A 5.00-kg body is raised 3.00 m in the earth's gravitational field. **(a)** How much work is done? **(b)** How much work would be done if the distance were doubled?

S1-9. Suppose that a perfectly round, homogeneous rubber ball falls freely along a line perpendicular to the earth's surface, its initial trajectory directed toward the center of the planet. Are there any circumstances under which the ball will bounce away at a 45° angle? Explain.

S1-10. Why do some balls bounce higher than others when released from the same height?

S1-11. Why does a ball bounce higher if it is *thrown* (rather than simply dropped) to the ground?

S1-12. Both a feather and a truck are released from a height of 100 m above the earth. Which object hits the ground first?

S1-13. The same feather and truck are released from a height of 100 m above the moon. **(a)** Which object hits the lunar surface first? **(b)** Which object delivers more momentum when it hits? **(c)** Which object is moving faster when it hits?

S1-14. A 1000-kg mass is dropped from a height of 10 m above the earth and also from a height of 10 m above the moon. Which surface—terrestrial or lunar—receives a larger dose of momentum?

S1-15. For each combination of units, state whether the quantity can possibly represent either velocity, acceleration, momentum, or energy:

(a) lb g^{-1} **(b)** lb mi h ft^{-1} **(c)** $\text{lb mi}^{3}\text{ ft}^{-1}\text{ h}^{-1}$ **(d)** lb min kg^{-1}

Note that the *pound* (lb) is a unit of force in the English system.

S1-16. Show that the statement below is dimensionally correct:

$$\text{Energy} \sim \frac{(\text{momentum})^2}{\text{mass}}$$

S1-17. What quantity is expressed in units of *foot-pounds* (ft lb)?

S1-18. Show that the statement below is dimensionally correct:

$$\text{Momentum} \sim \frac{\text{energy} \times \text{time}}{\text{length}}$$

S1-19. The *watt*, used to measure electric power, represents the quantity *work per unit time*. Express the watt in terms of kg, m, and s.

S1-20. The electric field, defined as force per unit charge, is expressed naturally as *newtons per coulomb*. Show that the electric field can also be stated in units of *volts per meter*.

S1-21. The acceleration due to gravity (g) is approximately 9.81 m s^{-2} near the surface of the earth, a body with a mass of 5.98×10^{27} g. Estimate the radius of the earth, assuming the planet to be a sphere. Note that the value of the universal gravitational constant (G) is 6.67×10^{-11} m^3 kg^{-1} s^{-2}.

S1-22. The moon has a radius of 1.74×10^6 m, and the acceleration due to lunar gravity is 1.62 m s^{-2}. Estimate the mass of the moon.

S1-23. Suppose that Planet X has a mass of 1.00×10^{26} kg and a radius of 1.00×10^8 m. Estimate the acceleration due to gravity near the surface of Planet X.

SOLUTIONS

S1-1. Each conversion is accomplished by stringing together the appropriate unit factors.

(a)
$$\frac{3.00 \times 10^8 \text{ m}}{\text{s}} \times \frac{1 \text{ km}}{1000 \text{ m}} \times \frac{60 \text{ s}}{\text{min}} = 1.80 \times 10^7 \text{ km min}^{-1}$$

(b)
$$\frac{3.00 \times 10^8 \text{ m}}{\text{s}} \times \frac{100 \text{ cm}}{\text{m}} \times \frac{60 \text{ s}}{\text{min}} \times \frac{60 \text{ min}}{\text{h}} \times \frac{24 \text{ h}}{\text{d}} \times \frac{7 \text{ d}}{\text{wk}} = 1.81 \times 10^{16} \text{ cm wk}^{-1}$$

(c)
$$\frac{3.00 \times 10^8 \text{ m}}{\text{s}} \times \frac{100 \text{ cm}}{\text{m}} \times \frac{1 \text{ in}}{2.54 \text{ cm}} \times \frac{1 \text{ ft}}{12 \text{ in}} \times \frac{1 \text{ mi}}{5280 \text{ ft}} \times \frac{3600 \text{ s}}{\text{h}} = 6.71 \times 10^8 \text{ mi h}^{-1}$$

(d)
$$\frac{3.00 \times 10^8 \text{ m}}{\text{s}} \times \frac{100 \text{ cm}}{\text{m}} \times \frac{1 \text{ in}}{2.54 \text{ cm}} \times \frac{3600 \text{ s}}{\text{h}} \times \frac{24 \text{ h}}{\text{d}} \times \frac{365.25 \text{ d}}{\text{y}} = 3.73 \times 10^{17} \text{ in y}^{-1}$$

S1-2. The relationship

$$1 \text{ L} = 1000 \text{ cm}^3$$

allows us to convert cubic centimeters into liters:

$$V = (5.00 \text{ cm})^3 \times \frac{1 \text{ L}}{1000 \text{ cm}^3} = 0.125 \text{ L}$$

Recall that the volume of a cube with edge l is equal to l^3.

S1-3. Use the formula for the volume of a sphere,

$$V = \frac{4}{3}\pi r^3$$

and then convert cubic meters into milliliters:

$$V = \frac{4}{3}\pi(1.00 \text{ m})^3 \times \left[\left(\frac{100 \text{ cm}}{\text{m}}\right)^3 \times \frac{1 \text{ mL}}{\text{cm}^3}\right] = 4.19 \times 10^6 \text{ mL}$$

S1-4. The defining relationship between centimeters and inches,

$$1 \text{ in} = 2.54 \text{ cm}$$

facilitates conversion between spatial measurements expressed in SI and English units. Note that significant figures are not relevant for exact quantities.

(a) $4840 \text{ yd}^2 \times \left(\frac{3 \text{ ft}}{\text{yd}}\right)^2 = 43{,}560 \text{ ft}^2$

(b) $4840 \text{ yd}^2 \times \left(\frac{3 \text{ ft}}{\text{yd}} \times \frac{12 \text{ in}}{\text{ft}} \times \frac{2.54 \text{ cm}}{\text{in}} \times \frac{1 \text{ m}}{100 \text{ cm}}\right)^2 = 4046.86 \text{ m}^2$

S1-5. We calculate the volume of a cube with edge equal to l,

$$= l^3$$

to obtain the number density per milliliter:

$$\frac{N}{V} = \frac{6.0 \times 10^{23} \text{ atoms}}{(28.2 \text{ cm})^3 \times \frac{1 \text{ mL}}{\text{cm}^3}} = 2.7 \times 10^{19} \text{ atoms mL}^{-1}$$

S1-6. The elementary charge on an electron is equal to -1.602×10^{-19} C:

$$-1.00 \text{ C} \times \frac{1 \text{ e}^-}{-1.602 \times 10^{-19} \text{ C}} = 6.24 \times 10^{18} \text{ e}^-$$

S1-7. Coulomb's law specifies the force F between electrically charged particles separated by a distance r:

$$F = \frac{1}{4\pi\varepsilon_0} \frac{q_1 q_2}{r^2} \qquad \left(\varepsilon_0 = 8.854 \times 10^{-12} \text{ C}^2 \text{ N}^{-1} \text{ m}^{-2}\right)$$

The symbols q_1 and q_2 represent the two charges, and the symbol ε_0 represents the permittivity of vacuum.

(a) Convert centimeters into meters to obtain the force in newtons:

$$F = \frac{1}{4\pi\left(8.854 \times 10^{-12} \text{ C}^2 \text{ N}^{-1} \text{ m}^{-2}\right)} \frac{(-2.00 \text{ C})(-2.00 \text{ C})}{\left(1.00 \text{ cm} \times \dfrac{1 \text{ m}}{100 \text{ cm}}\right)^2} = 3.60 \times 10^{14} \text{ N}$$

The force between like charges is repulsive. Its sign is positive.

(b) Proportional to r^{-2}, the force falls by a factor of 16 (equivalently, 6.25%) when the separation is quadrupled (here, from $r = 1.00$ cm to $r' = 4r = 4.00$ cm):

$$\frac{F(4r)}{F(r)} = \frac{r^2}{(4r)^2} = \frac{1}{16}$$

(c) Applying the same reasoning, we show that the force is tripled when the distance is scaled by $1/\sqrt{3}$:

$$\frac{F(r/\sqrt{3})}{F(r)} = \frac{r^2}{\left(\dfrac{r}{\sqrt{3}}\right)^2} = 3$$

With r originally equal to 1.00 cm, the corresponding value $r/\sqrt{3}$ is 0.577 cm.

S1-8. The work done by the force of gravity,

$$W = -F\,\Delta h = -mg\,\Delta h \qquad (g = 9.81 \text{ m s}^{-2})$$

is taken as negative for a positive displacement Δh.

(a) Substitute 5.00 kg for the mass and 3.00 m for the displacement:

$$W = -mg\,\Delta h = -(5.00 \text{ kg})(9.81 \text{ m s}^{-2})(3.00 \text{ m}) = -147 \text{ kg m}^2 \text{ s}^{-2} \equiv -147 \text{ J}$$

(b) The work, directly proportional to Δh, doubles when the displacement is doubled.

S1-9. Were the ball to rebound at a 45° angle, the process would violate the law of momentum conservation. There is no initial component parallel to the surface of the earth:

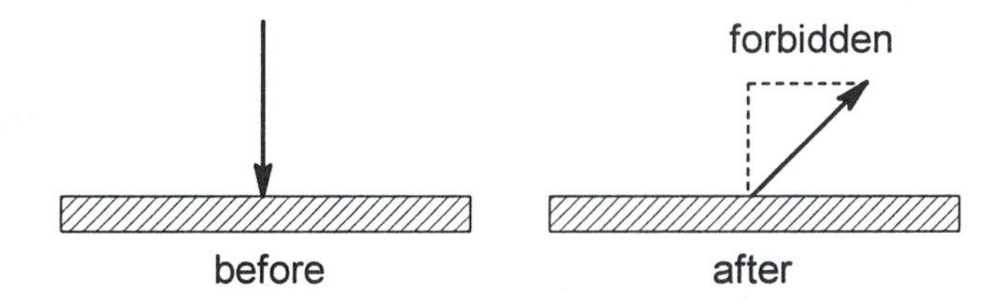

S1-10. No collision is truly elastic. Any *real* ball (as opposed to an ideal ball) will interact inelastically with the ground, dissipating a certain amount of energy in the form of heat. A portion of the total kinetic energy is transferred into random molecular motion, and the ball is unable to rise to its original height. The height of the rebound depends on the material composition of the object.

S1-11. A thrown ball, moving with an initial velocity v_0, has a nonzero initial *kinetic* energy

$$E_{\text{k0}} = \frac{1}{2}m(v_0)^2$$

in addition to its gravitational potential energy mgh_0. This fixed amount of energy,

$$E_{\text{tot}} = mgh_0 + \frac{1}{2}m(v_0)^2$$

is later converted wholly into gravitational potential energy when the bouncing ball momentarily stops at a new height h greater than h_0:

$$mgh = mgh_0 + \frac{1}{2}m(v_0)^2$$

$$h = h_0 + \frac{(v_0)^2}{2g} > h_0$$

By contrast, a freely falling ball ($v_0 = 0$) has only its initial potential energy and cannot rebound past its initial height.

S1-12. In vacuum, the two objects would hit the ground simultaneously since acceleration in a gravitational field is independent of mass. Resistance by the earth's atmosphere, however, slows the fall of the feather. The truck hits first.

S1-13. See the previous exercise.

(a) In the vacuum of space, both objects hit the lunar surface at the same instant and with the same velocity.

(b) The more massive object, the truck, delivers more momentum upon impact:

$$p = mv$$

(c) The two bodies move at the same speed, since acceleration in a gravitational field is independent of mass.

S1-14. The terrestrial surface absorbs more momentum. The acceleration of gravity is six times greater on the earth, and the falling mass attains a correspondingly higher terminal velocity.

S1-15. The named quantities all reduce to fundamental dimensions of mass, length, and time:

$$\text{Velocity} \sim \frac{\text{length}}{\text{time}}$$

$$\text{Acceleration} \sim \frac{\text{velocity}}{\text{time}} \sim \frac{\text{length}}{(\text{time})^2}$$

$$\text{Momentum} \sim \text{mass} \times \text{velocity} \sim \frac{\text{mass} \times \text{length}}{\text{time}}$$

$$\text{Energy} \sim \text{force} \times \text{length} \sim \frac{(\text{momentum})^2}{\text{mass}} \sim \text{mass} \times (\text{velocity})^2 \sim \frac{\text{mass} \times (\text{length})^2}{(\text{time})^2}$$

We interpret the English-system *pound* (lb) as a unit of force.

(a) Newton's second law states that force is equal to mass × acceleration:

$$\frac{\text{lb}}{\text{g}} \sim \frac{\text{force}}{\text{mass}} = \text{acceleration}$$

(b) Momentum is equivalently expressed as mass × velocity (units = kg m s^{-1}) or force × time (units = kg m s^{-2} s = kg m s^{-1}):

$$\frac{\text{lb mi h}}{\text{ft}} \sim \frac{\text{force} \times \text{length} \times \text{time}}{\text{length}} = \text{force} \times \text{time} \sim \text{momentum}$$

(c) The reduced dimensions correspond to neither velocity nor acceleration nor momentum nor energy:

$$\frac{\text{lb mi}^3}{\text{ft h}} \sim \frac{\text{force} \times (\text{length})^3}{\text{length} \times \text{time}} = \frac{\text{force} \times (\text{length})^2}{\text{time}}$$

Instead, the quantity may represent energy × velocity:

$$\frac{\text{force} \times (\text{length})^2}{\text{time}} = (\text{force} \times \text{length}) \times \frac{\text{length}}{\text{time}} \sim \text{energy} \times \text{velocity}$$

(d) The units reduce to dimensions of velocity (length/time):

$$\frac{\text{lb min}}{\text{kg}} \sim \frac{\text{force} \times \text{time}}{\text{mass}} = \text{acceleration} \times \text{time} = \frac{\text{length}}{(\text{time})^2} \times \text{time} \sim \text{velocity}$$

S1-16. The expression has dimensions of kinetic energy (proportional to mv^2):

$$\frac{(\text{momentum})^2}{\text{mass}} \sim \frac{(\text{mass} \times \text{velocity})^2}{\text{mass}} = \text{mass} \times (\text{velocity})^2$$

Inserting the SI base units kg, m, and s,

$$\text{mass} \times (\text{velocity})^2 \sim \text{kg m}^2\ \text{s}^{-2} = \text{J}$$

we recover the *joule*, the SI derived unit of work and energy.

S1-17. The foot-pound, a product of length and force, has dimensions of work (energy):

$$\text{ft lb} \sim \text{length} \times \text{force} = \text{work}$$

S1-18. Substitute, for example, the dimensional relationships

$$\text{Energy} \sim \text{mass} \times (\text{velocity})^2$$

$$\text{Velocity} \sim \frac{\text{length}}{\text{time}}$$

to obtain dimensions of momentum (mass × velocity):

$$\frac{\text{energy} \times \text{time}}{\text{length}} \sim \frac{\text{mass} \times (\text{velocity})^2 \times \text{time}}{\text{length}} = \frac{\text{mass} \times (\text{velocity})^2}{\text{velocity}} = \text{mass} \times \text{velocity}$$

In SI units, the resulting quantity is represented as kg m s^{-1}:

$$\frac{\text{energy} \times \text{time}}{\text{length}} \sim \frac{\text{J s}}{\text{m}} = \frac{\text{kg m}^2\ \text{s}^{-2}\ \text{s}}{\text{m}} = \text{kg m s}^{-1}$$

Equivalent expressions of momentum (such as force × time) carry the same units, as they must.

S1-19. Use the definition 1 J = 1 kg $\text{m}^2\ \text{s}^{-2}$:

$$\text{Watt} \sim \frac{\text{work}}{\text{time}} \sim \frac{\text{J}}{\text{s}} = \frac{\text{kg m}^2\ \text{s}^{-2}}{\text{s}} = \text{kg m}^2\ \text{s}^{-3}$$

S1-20. Use the definition 1 V = 1 J C^{-1}:

$$\frac{\text{N}}{\text{C}} = \frac{\text{N m}}{\text{C m}} = \frac{\text{J}}{\text{C m}} = \frac{\text{V}}{\text{m}}$$

Alternatively, reduce both expressions to their fundamental dimensions of mass, length, time, and charge:

$$\frac{\text{N}}{\text{C}} = \frac{\text{kg m s}^{-2}}{\text{C}}$$

$$\frac{\text{V}}{\text{m}} = \frac{\text{J}}{\text{C m}} = \frac{\text{kg m}^2\ \text{s}^{-2}}{\text{C m}} = \frac{\text{kg m s}^{-2}}{\text{C}}$$

S1-21. Treat the earth as a perfect sphere with radius R and mass M. A test mass m on the surface experiences a gravitational force of magnitude

$$|F| = \frac{GMm}{R^2}$$

and a corresponding acceleration g:

$$g = \frac{|F|}{m} = \frac{GM}{R^2}$$

Given the values $G = 6.67 \times 10^{-11}\ \text{m}^3\ \text{kg}^{-1}\ \text{s}^{-2}$, $M = 5.98 \times 10^{27}$ g, and $g = 9.81\ \text{m s}^{-2}$, we have sufficient information to calculate the radius R:

$$R = \sqrt{\frac{GM}{g}} = \sqrt{\frac{\left(6.67 \times 10^{-11}\ \text{m}^3\ \text{kg}^{-1}\ \text{s}^{-2}\right)\left(5.98 \times 10^{27}\ \text{g} \times \frac{1\ \text{kg}}{1000\ \text{g}}\right)}{9.81\ \text{m s}^{-2}}} = 6.38 \times 10^6\ \text{m}$$

S1-22. See the preceding exercise for the method in general. Here, given R and g for the moon, we solve for the lunar mass M:

$$M = \frac{gR^2}{G} = \frac{\left(1.62\ \text{m s}^{-2}\right)\left(1.74 \times 10^6\ \text{m}\right)^2}{6.67 \times 10^{-11}\ \text{m}^3\ \text{kg}^{-1}\ \text{s}^{-2}} = 7.35 \times 10^{22}\ \text{kg}$$

S1-23. From Exercise S1-21, we know how to calculate the gravitational acceleration near the surface of a spherical planet:

$$g = \frac{GM}{R^2} = \frac{\left(6.67 \times 10^{-11}\ \text{m}^3\ \text{kg}^{-1}\ \text{s}^{-2}\right)\left(1.00 \times 10^{26}\ \text{kg}\right)}{\left(1.00 \times 10^8\ \text{m}\right)^2} = 0.667\ \text{m s}^{-2}$$

Chapter 2

Atoms and Molecules

S2-1. How many protons, neutrons, and electrons are contained in each of the following isotopes?

(a) ^{33}S **(b)** ^{34}S **(c)** ^{36}S

S2-2. How many protons, neutrons, and electrons are contained in each of the following isotopes?

(a) ^{28}Si **(b)** ^{29}Si **(c)** ^{30}Si

S2-3. A certain element X has four isotopes: ^{54}X, ^{56}X, ^{57}X , ^{58}X. The ion X^{3+} contains a total of 23 electrons. **(a)** Identify X. **(b)** State the number of protons and neutrons in each isotope.

S2-4. Chromium exists as four isotopes: ^{50}Cr, ^{52}Cr, ^{53}Cr, ^{54}Cr. Use the information in the table below to determine the molar mass of chromium in its natural state.

	MOLAR MASS ($g\ mol^{-1}$)	ABUNDANCE (%)
^{50}Cr	49.946046	4.345
^{52}Cr	51.940509	83.789
^{53}Cr	52.940651	9.501
^{54}Cr	53.938882	2.365

S2-5. The element boron exists as two isotopes: ^{10}B, ^{11}B. Use the information in the table below to determine the molar mass of ^{10}B.

	MOLAR MASS (g mol^{-1})	ABUNDANCE (%)
^{10}B	—	19.9
^{11}B	11.009305	80.1

Consult the periodic table as needed.

S2-6. How many core electrons and valence electrons are contained in each of the following atoms and ions?

(a) Ca **(b)** Ti^{2+} **(c)** Ba **(d)** Fr **(e)** Al

S2-7. How many core electrons and valence electrons are contained in each of the following atoms?

(a) N **(b)** P **(c)** O **(d)** S **(e)** F **(f)** Cl

S2-8. How many core electrons and valence electrons are contained in each of the following molecules?

(a) N_2 **(b)** O_2 **(c)** F_2 **(d)** Cl_2

S2-9. Which of the following atoms are likely to form anions? Which are likely to form cations?

(a) Cs **(b)** O **(c)** Cr **(d)** I **(e)** Ra

S2-10. A certain molecule consists of one atom of carbon and four atoms of iodine. **(a)** Draw the Lewis structure. **(b)** Use the VSEPR model to predict the most likely geometry.

S2-11. A certain molecule consists of one atom of beryllium and two atoms of chlorine. **(a)** Draw the Lewis structure. **(b)** Use the VSEPR model to predict the most likely geometry.

S2-12. **(a)** Draw a Lewis structure for the molecule GeF_2. Assume that germanium contributes four valence electrons. **(b)** According to the VSEPR model, how many valence electrons are present on the central atom? How are they arranged? **(c)** What is the expected molecular geometry of GeF_2?

S2-13. **(a)** Draw a Lewis structure for the molecule ClF_3. **(b)** According to the VSEPR model, how many pairs of valence electrons are present on the central atom? How are they arranged? **(c)** What is the expected molecular geometry?

S2-14. **(a)** Draw a Lewis structure for the molecular ion ICl_2^-. Assume that the iodine atom contributes seven valence electrons. **(b)** According to the VSEPR model, how many pairs of valence electrons are present on the central atom? How are they arranged? **(c)** What is the expected molecular geometry?

S2-15. **(a)** Draw a Lewis structure for the molecular ion $SnCl_3^-$. Assume that tin contributes four valence electrons. **(b)** According to the VSEPR model, how many pairs of valence electrons are present on the central atom? How are they arranged? **(c)** What is the expected molecular geometry?

S2-16. If the price of gold is $300 per troy ounce (31.1 g), what is the value of a single atom?

S2-17. If the price of silver, expressed in British pounds sterling, is £3.00 per troy ounce (31.1 g), what is the equivalent price in dollars per mole? Assume an exchange rate of £1.00 = $1.60.

S2-18. How many atoms of copper are there in a metric ton of the metal (1000 kg)?

S2-19. The Milky Way contains on the order of 100 billion stars. **(a)** Approximately how many "moles of stars" are present in our galaxy? **(b)** If each star had the mass of a hydrogen atom, what would be the total mass of the Milky Way?

S2-20. A Robber Baron exploits the masses at the rate of 1 million dollars per hour. Approximately how many years will the Robber Baron need to accumulate a fortune equal to 1 mole of dollars?

S2-21. Calculate the percent composition of each element in the following compounds:

(a) $C_6H_{12}O_6$ **(b)** Al_2O_3 **(c)** Ag_2SO_4 **(d)** Na_2Te

Which of the compounds are molecular and which are ionic?

S2-22. Calculate the percent composition of each element in the following compounds:

(a) N_2O_5 **(b)** C_6H_5OH **(c)** CH_3COOH **(d)** $ZnGa_2O_4$

Which of the compounds are molecular and which are ionic?

S2-23. Use the elemental analysis given below to determine an empirical formula:

Na: 18.78% Cl: 28.95% O: 52.27%

S2-24. The molar mass of an unknown molecular compound is 56.11 g mol^{-1}. **(a)** Use the elemental data tabulated below to determine an empirical formula:

C: 85.63% H: 14.37%

(b) What is the molecular formula?

S2-25. Balance the following equation:

$$\_\_ HNO_3 \rightarrow \_\_ H_2 + \_\_ N_2 + \_\_ O_2$$

S2-26. Balance the following equation:

$$\_\_ NH_3 + \_\_ O_2 \rightarrow \_\_ NO + \_\_ H_2O$$

S2-27. Consider the reaction below:

$$BF_3 + NH_3 \rightarrow BF_3NH_3$$

(a) How many grams of BF_3 will react completely with 50.00 g NH_3? **(b)** How many grams of BF_3NH_3 will be produced as a result?

S2-28. The gas-phase reaction of ozone and nitric oxide is described by the following chemical equation:

$$O_3 + NO \rightarrow NO_2 + O_2$$

(a) Suppose that equal masses of O_3 and NO are present. Which of the two reactants is stoichiometrically limiting? **(b)** How many grams of NO_2 and O_2 are produced when 10.00 g NO are completely consumed?

S2-29. Nitric oxide reacts with molecular bromine according to the following chemical equation:

$$2NO + Br_2 \rightarrow 2NOBr$$

(a) How many grams of NOBr are produced when 50.00 g NO react with 50.00 g Br_2? **(b)** Which of the two reactants is consumed completely? **(c)** How many grams of the other reactant remain unconsumed?

SOLUTIONS

S2-1. Ascertain the atomic number of the element (Z, the number of protons) from its chemical symbol. Then, from the superscript in each isotopic symbol ^{A}X, identify the *mass number A* (the combined number of protons and neutrons). Next, solve for the *neutron number*

$$N = A - Z$$

by simple rearrangement of the defining equation for A:

$$Z + N = A$$

Isotopes of a given element have the same number of protons but different numbers of neutrons—and, consequently, different mass numbers.

Finally, observe that the total number of electrons (N_e) in a neutral atom is identical to Z, the number of protons in its nucleus.

	^{A}X	A	Z	N	N_e
(a)	^{33}S	33	16	17	16
(b)	^{34}S	34	16	18	16
(c)	^{36}S	36	16	20	16

S2-2. The method is the same as in the preceding exercise.

	^{A}X	A	Z	N	N_e
(a)	^{28}Si	28	14	14	14
(b)	^{29}Si	29	14	15	14
(c)	^{30}Si	30	14	16	14

S2-3. An element derives its identity from its atomic number Z, the number of protons. In a neutral atom, the positive nuclear charge ($+Z$) is balanced by exactly Z electrons outside the nucleus. The number of electrons falls to $Z - q$ in a monatomic cation with charge q, and it rises to $Z + |q|$ in a monatomic anion with charge $-q$.

(a) Since X^{3+} contains 23 electrons, the neutral atom must contain 26. The atomic number (Z) is therefore 26, and the element is iron: Fe.

(b) The mass number A is equal to the sum of the atomic number Z and the neutron number N:

^{A}X	A	Z	N
^{54}Fe	54	26	28
^{56}Fe	56	26	30
^{57}Fe	57	26	31
^{58}Fe	58	26	32

S2-4. The average molar mass $\mathcal{m}$ arises as a weighted average over the four isotopic masses,

$$\mathcal{m} = X_{50}m_{50} + X_{52}m_{52} + X_{53}m_{53} + X_{54}m_{54}$$

where X_A and m_A denote, respectively, the fractional abundance and molar mass of the isotope ^{A}X. We simply insert the given values and evaluate the resulting expression:

$$\mathcal{m} = (0.04345 \times 49.946046 + 0.83789 \times 51.940509 + 0.09501 \times 52.940651$$

$$+ 0.02365 \times 53.938882)\text{ g mol}^{-1} = 51.996\text{ g mol}^{-1}$$

S2-5. Follow the same reasoning as in the preceding exercise, using the equation

$$\mathcal{m} = X_{10}m_{10} + X_{11}m_{11}$$

to solve for one of the contributing isotopic masses:

$$m_{10} = \frac{\mathcal{m} - X_{11}m_{11}}{X_{10}} = \left(\frac{10.811 - 0.801 \times 11.009305}{0.199}\right)\text{ g mol}^{-1} = 10.013\text{ g mol}^{-1}$$

Rounded to three significant figures, the value is 10.0 g mol^{-1}.

S2-6. First, we recall that the total number of electrons in any monatomic species X^{q} is equal to $Z - q$ (where q is a signed number):

SPECIES	CHARGE	N_e
monatomic cation	X^{q+}	$Z - q$
neutral atom	X	Z
monatomic anion	X^{q-}	$Z + \lvert q \rvert$

Next, of the total complement of N_e electrons, we assign N_{core} to the core and N_{val} to the valence:

$$N_e = N_{core} + N_{val}$$

In the examples below, the number of core electrons—the number of electrons in fully occupied shells—is equal to the atomic number of the noble gas in the row last completed. Finally, having established both N_e and N_{core}, we can solve for the number of valence electrons by subtraction:

$$N_{val} = N_e - N_{core}$$

Results are summarized in the following table:

	X^q	Z	q	$N_e = Z - q$	N_{core}	$N_{val} = N_e - N_{core}$
(a)	Ca	20	0	20	18	2
(b)	Ti^{2+}	22	+2	20	18	2
(c)	Ba	56	0	56	54	2
(d)	Fr	87	0	87	86	1
(e)	Al	13	0	13	10	3

S2-7. Similar, although here we have only neutral atoms to consider:

	X	Z	$N_e = Z$	N_{core}	$N_{val} = N_e - N_{core}$
(a)	N	7	7	2	5
(b)	P	15	15	10	5
(c)	O	8	8	2	6
(d)	S	16	16	10	6
(e)	F	9	9	2	7
(f)	Cl	17	17	10	7

S2-8. Similar again, but now for diatomic molecules. The total number of electrons in a neutral species is equal to the sum of the atomic numbers. Every atom, in turn, contributes its core electrons and valence electrons to the corporate structure:

	X_2	Z	$N_e = 2Z$	N_{core}	$N_{val} = N_e - N_{core}$
(a)	N_2	7	14	4	10
(b)	O_2	8	16	4	12
(c)	F_2	9	18	4	14
(d)	Cl_2	17	34	20	14

See the preceding exercise for the determination of N_{core} and N_{val} for each atom.

S2-9. Metals, species with low electronegativities, tend to lose electrons and thus form cations. Nonmetals, which gain electrons, typically form anions.

	ATOM	CLASSIFICATION	ION
(a)	Cs	alkali metal	cation
(b)	O	chalcogen (nonmetal)	anion
(c)	Cr	transition metal	cation
(d)	I	halogen (nonmetal)	anion
(e)	Ra	alkaline earth metal	cation

Electronegativity decreases toward the bottom and left of the periodic table. Electron affinity (the tendency to gain an electron) increases toward the top and right.

S2-10. The carbon atom contributes four valence electrons, and each of the four iodine atoms contributes seven. There are 32 valence electrons in all.

(a) The 32 valence electrons are distributed as follows:

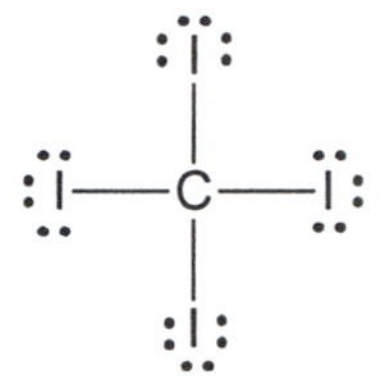

I—C—I (with I above and below C; each I with three lone pairs)

Carbon attains an octet by forming four single bonds, leaving each iodine site with three lone pairs.

For brevity, we often represent such a structure with the lone pairs suppressed:

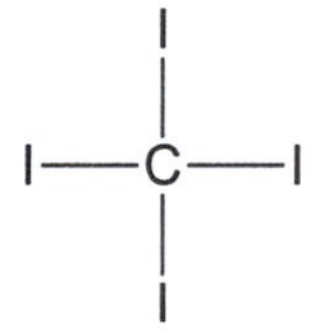

I—C—I (with I above and below C)

(b) The VSEPR model predicts a tetrahedral geometry around the central carbon atom, both for the four electron pairs and for their attached atoms.

S2-11. $BeCl_2$ contains 16 valence electrons: two from beryllium and seven apiece from each chlorine.

(a) Beryllium, with two bonding pairs and no lone pairs, remains four electrons short of an octet:

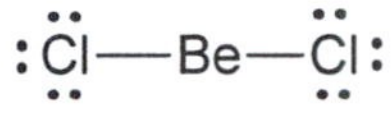

:Cl—Be—Cl: (each Cl with three lone pairs)

(b) The molecule is expected to be linear. VSEPR theory predicts that the two electron pairs on Be will point in opposite directions, creating a 180° Cl–Be–Cl bond angle.

S2-12. Germanium, in Group IV, supplies four valence electrons. Each of the two fluorine atoms contributes seven, making a total of 18 electrons to distribute.

(a) A plausible Lewis structure, albeit one with an octet-deficient germanium atom, is shown below:

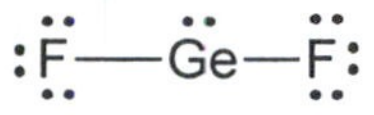

:F—Ge—F:

(b) Two bonding pairs and one lone pair—a total of six electrons—surround the central germanium atom. Electrostatic repulsions are minimized when the three electron pairs point to the vertices of an equilateral triangle.

(c) The molecule is expected to be bent, with a bond angle of approximately 120°:

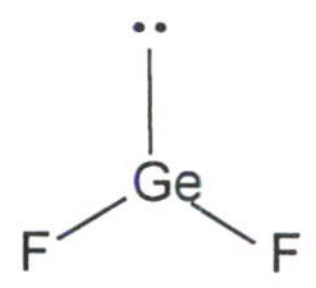

Ge
F F

The lone pair, occupying one of the three vertices, has no atom attached.

S2-13. We have 28 valence electrons to distribute, seven from each halogen atom in ClF_3.

(a) Chlorine exhibits an expanded "octet" of 10 electrons:

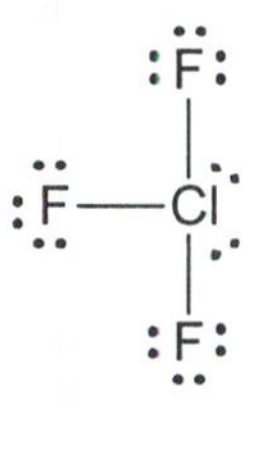

:F:
:F—Cl
:F:

(b) Five pairs of valence electrons are arranged about the central chlorine atom: three bonding pairs and two lone pairs. The five electron pairs are oriented toward the vertices of a trigonal bipyramid.

(c) The molecule is T-shaped. Two positions in the equatorial plane of the trigonal bipyramid are occupied by lone pairs (not atoms):

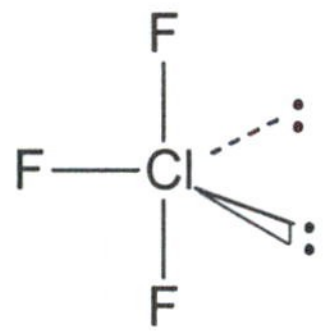

F
F—Cl
F

S2-14. There are 22 valence electrons in ICl_2^- : seven from each of the three halogen atoms, plus one extra for the single negative charge.

(a) The central iodine atom accommodates 10 electrons in an expanded octet:

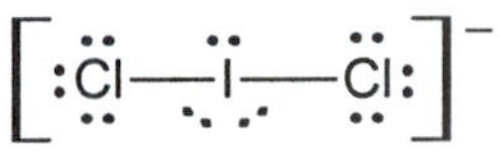

(b) Five electron pairs on iodine (two bonding pairs, three lone pairs) point to the five vertices of a trigonal bipyramid.

(c) The ion is linear. The three positions in the equatorial plane are occupied by lone pairs, not atoms:

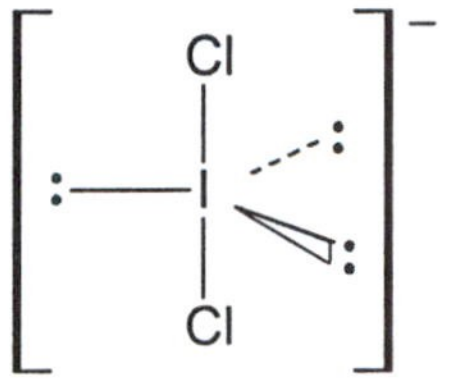

S2-15. Tin, in Group IV, supplies four valence electrons. Each of three chlorine atoms contributes seven, and one extra electron leaves $SnCl_3^-$ with a net charge of -1.

(a) All four atoms have an octet in the following 26-electron Lewis structure:

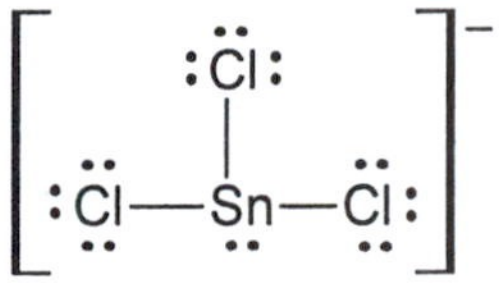

(b) There are four pairs of electrons on the central Sn site: three bonding pairs and one lone pair. They point to the four vertices of a tetrahedron.

(c) The four atoms are arranged as a trigonal pyramid, with Sn at the apex:

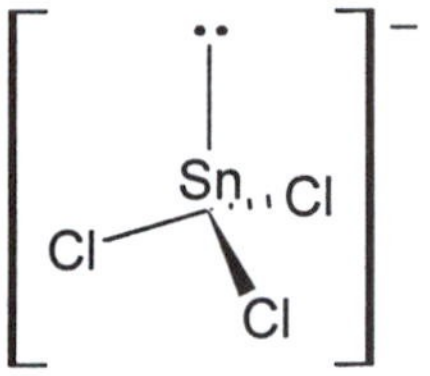

The fourth vertex of the electronic tetrahedron, containing a lone pair, has no attached atom.

S2-16. A straightforward conversion of grams into moles into particles:

$$\frac{\$300}{31.1 \text{ g Au}} \times \frac{196.967 \text{ g Au}}{\text{mol}} \times \frac{1 \text{ mol}}{6.022 \times 10^{23} \text{ atoms}} = \$3.16 \times 10^{-21} \text{ per atom}$$

S2-17. Similar to the preceding exercise:

$$\frac{£3.00}{31.1 \text{ g Ag}} \times \frac{107.868 \text{ g Ag}}{\text{mol}} \times \frac{\$1.60}{£1.00} = \$16.65 \text{ mol}^{-1}$$

S2-18. Use the molar mass and Avogadro's number to go from kilograms to moles to atoms:

$$1000 \text{ kg Cu} \times \frac{1000 \text{ g}}{\text{kg}} \times \frac{1 \text{ mol}}{63.546 \text{ g Cu}} \times \frac{6.022 \times 10^{23} \text{ atoms}}{\text{mol}} = 9.477 \times 10^{27} \text{ atoms}$$

S2-19. Even astronomical numbers pale in comparison to the number of microscopic particles in one mole.

(a) One hundred billion particles amounts to scarcely more than 0.1 trillionth mole:

$$10^{11} \text{ stars} \times \frac{1 \text{ mol}}{6.02 \times 10^{23} \text{ stars}} = 1.66 \times 10^{-13} \text{ mol}$$

(b) Again, the corresponding microscopic quantity proves to be infinitesimally small:

$$10^{11} \text{ particles} \times \frac{1 \text{ mol}}{6.02 \times 10^{23} \text{ particles}} \times \frac{1.00794 \text{ g}}{\text{mol}} = 1.67 \times 10^{-13} \text{ g}$$

S2-20. Avogadro's number is a *very* large amount:

$$6.02 \times 10^{23} \text{ dollars} \times \frac{1 \text{ h}}{10^6 \text{ dollars}} \times \frac{1 \text{ d}}{24 \text{ h}} \times \frac{1 \text{ y}}{365.25 \text{ d}} = 6.87 \times 10^{13} \text{ y}$$

S2-21. One sample calculation, for $C_6H_{12}O_6$, should suffice:

$$\frac{6 \text{ mol C}}{\text{mol } C_6H_{12}O_6} \times \frac{12.011 \text{ g C}}{\text{mol C}} \times \frac{1 \text{ mol } C_6H_{12}O_6}{180.158 \text{ g } C_6H_{12}O_6} \times 100\% = 40.002\% \text{ C} \quad \text{(mass)}$$

$$\frac{12 \text{ mol H}}{\text{mol } C_6H_{12}O_6} \times \frac{1.00794 \text{ g H}}{\text{mol H}} \times \frac{1 \text{ mol } C_6H_{12}O_6}{180.158 \text{ g } C_6H_{12}O_6} \times 100\% = 6.71371\% \text{ H} \quad \text{(mass)}$$

$$\frac{6 \text{ mol O}}{\text{mol } C_6H_{12}O_6} \times \frac{15.9994 \text{ g O}}{\text{mol O}} \times \frac{1 \text{ mol } C_6H_{12}O_6}{180.158 \text{ g } C_6H_{12}O_6} \times 100\% = 53.2846\% \text{ O} \quad \text{(mass)}$$

Brief results are summarized below.

(a) $C_6H_{12}O_6$, a carbohydrate, is a molecular compound:

$$\mathcal{M} = 180.158 \text{ g mol}^{-1} \qquad 40.002\% \text{ C} \qquad 6.71371\% \text{ H} \qquad 53.2846\% \text{ O}$$

(b) Al_2O_3 is an ionic compound:

$$\mathcal{M} = 101.9613 \text{ g mol}^{-1} \qquad 52.92507\% \text{ Al} \qquad 47.0749\% \text{ O}$$

(c) Ag_2SO_4 is an ionic compound consisting of Ag^+ cations and SO_4^{2-} anions:

$$\mathcal{M} = 311.800 \text{ g mol}^{-1} \qquad 69.1906\% \text{ Ag} \qquad 10.284\% \text{ S} \qquad 20.5252\% \text{ O}$$

(d) Na_2Te is also an ionic compound:

$$\mathcal{M} = 173.58 \text{ g mol}^{-1} \qquad 26.489\% \text{ Na} \qquad 73.511\% \text{ Te}$$

S2-22. The calculations are done the same way as in the preceding exercise.

(a) N_2O_5 is a molecular compound:

$$\mathcal{M} = 108.0104 \text{ g mol}^{-1} \qquad 25.9358\% \text{ N} \qquad 74.0642\% \text{ O}$$

(b) C_6H_5OH is a molecular compound:

$$\mathcal{M} = 94.113 \text{ g mol}^{-1} \qquad 76.574\% \text{ C} \qquad 6.4259\% \text{ H} \qquad 17.000\% \text{ O}$$

(c) CH_3COOH is a molecular compound:

$$\mathcal{M} = 60.053 \text{ g mol}^{-1} \qquad 40.002\% \text{ C} \qquad 6.7137\% \text{ H} \qquad 53.285\% \text{ O}$$

(d) $ZnGa_2O_4$ is an ionic compound:

$$\mathcal{M} = 268.83 \text{ g mol}^{-1} \qquad 24.32\% \text{ Zn} \qquad 51.871\% \text{ Ga} \qquad 23.806\% \text{ O}$$

S2-23. Assume that we have a sample containing 100 g:

$$18.78 \text{ g Na} \times \frac{1 \text{ mol Na}}{22.98977 \text{ g Na}} = 0.8169 \text{ mol Na}$$

$$28.95 \text{ g Cl} \times \frac{1 \text{ mol Cl}}{35.453 \text{ g Cl}} = 0.8166 \text{ mol Cl}$$

$$52.27 \text{ g O} \times \frac{1 \text{ mol O}}{15.9994 \text{ g O}} = 3.267 \text{ mol O}$$

Reduced to smallest integers, the molar proportions in $Na_{0.8169}Cl_{0.8166}O_{3.267}$ yield the empirical formula $NaClO_4$:

$$Na_{\frac{0.8169}{0.8166}}Cl_{\frac{0.8166}{0.8166}}O_{\frac{3.267}{0.8166}} = Na_{1.000}Cl_{1.000}O_{4.001}$$

S2-24. The method is the same as in the preceding exercise.

(a) Assume 100 g of sample:

$$85.63 \text{ g C} \times \frac{1 \text{ mol C}}{12.011 \text{ g C}} = 7.129 \text{ mol C}$$

$$14.37 \text{ g H} \times \frac{1 \text{ mol H}}{1.00794 \text{ g H}} = 14.26 \text{ mol H}$$

The empirical formula is CH_2.

(b) CH_2 has a formula weight of 14.027 g mol^{-1}:

$$\frac{1 \text{ mol C}}{\text{mol CH}_2} \times \frac{12.011 \text{ g}}{\text{mol C}} + \frac{2 \text{ mol H}}{\text{mol CH}_2} \times \frac{1.00794 \text{ g}}{\text{mol H}} = \frac{14.027 \text{ g}}{\text{mol CH}_2}$$

Since the molecular weight is 56.11 g mol^{-1}, we know that each molecule contains four CH_2 formula units. The molecular formula is C_4H_8.

S2-25. The equation is balanced by inspection, one step at a time:

$$\_\_HNO_3 \rightarrow \_\_H_2 + \_\_N_2 + \_\_O_2 \qquad \text{(unbalanced)}$$

$$\mathbf{2}HNO_3 \rightarrow H_2 + N_2 + \_\_O_2 \qquad \text{(H and N are balanced)}$$

$$2HNO_3 \rightarrow H_2 + N_2 + \mathbf{3}O_2 \qquad \text{(O is balanced)}$$

S2-26. Similar:

$$\_\_NH_3 + \_\_O_2 \rightarrow \_\_NO + \_\_H_2O \qquad \text{(N is balanced)}$$

$$NH_3 + \_\_O_2 \rightarrow NO + \tfrac{3}{2}H_2O \qquad \text{(H is balanced)}$$

$$NH_3 + \tfrac{5}{4}O_2 \rightarrow NO + \tfrac{3}{2}H_2O \qquad \text{(O is balanced)}$$

Multiplying through by 4 (which is not required), we obtain an equation expressed in whole numbers:

$$4NH_3 + 5O_2 \rightarrow 4NO + 6H_2O$$

S2-27. The process, discussed at length in the text, is an acid–base reaction in the Lewis sense:

$$BF_3 + :NH_3 \rightarrow BF_3NH_3$$

(a) One mole of BF_3 reacts with one mole of NH_3:

$$50.00 \text{ g } NH_3 \times \frac{1 \text{ mol } NH_3}{17.0305 \text{ g } NH_3} \times \frac{1 \text{ mol } BF_3}{\text{mol } NH_3} \times \frac{67.806 \text{ g } BF_3}{\text{mol } BF_3} = 199.1 \text{ g } BF_3$$

(b) Similarly, one mole of BF_3NH_3 is produced per mole of NH_3:

$$50.00 \text{ g } NH_3 \times \frac{1 \text{ mol } NH_3}{17.0305 \text{ g } NH_3} \times \frac{1 \text{ mol } BF_3NH_3}{\text{mol } NH_3} \times \frac{84.837 \text{ g } BF_3NH_3}{\text{mol } BF_3NH_3} = 249.1 \text{ g } BF_3NH_3$$

S2-28. All four species react in 1:1 stoichiometry:

$$O_3 + NO \rightarrow NO_2 + O_2$$

(a) Assume 100 g of each reactant:

$$100 \text{ g } O_3 \times \frac{1 \text{ mol } O_3}{47.998 \text{ g } O_3} = 2.083 \text{ mol } O_3$$

$$100 \text{ g NO} \times \frac{1 \text{ mol NO}}{30.006 \text{ g NO}} = 3.333 \text{ mol NO}$$

O_3 is the limiting species in this 1:1 reaction. NO is present in excess.

(b) Convert grams of reactant into moles of reactant. Then convert moles of reactant into moles of product and grams of product:

$$10.00 \text{ g NO} \times \frac{1 \text{ mol NO}}{30.006 \text{ g NO}} \times \frac{1 \text{ mol } NO_2}{\text{mol NO}} \times \frac{46.006 \text{ g } NO_2}{\text{mol } NO_2} = 15.33 \text{ g } NO_2$$

$$10.00 \text{ g NO} \times \frac{1 \text{ mol NO}}{30.006 \text{ g NO}} \times \frac{1 \text{ mol } O_2}{\text{mol NO}} \times \frac{31.999 \text{ g } O_2}{\text{mol } O_2} = 10.66 \text{ g } O_2$$

S2-29. Two moles of NO react with one mole of Br_2 to produce two moles of NOBr:

$$2NO + Br_2 \rightarrow 2NOBr$$

(a) Determine, first, the initial molar amount of each reactant:

$$50.00 \text{ g NO} \times \frac{1 \text{ mol NO}}{30.006 \text{ g NO}} = 1.666 \text{ mol NO}$$

$$50.00 \text{ g Br}_2 \times \frac{1 \text{ mol Br}_2}{159.808 \text{ g Br}_2} = 0.3129 \text{ mol Br}_2$$

Br_2 is the limiting reactant. Constrained by 2:1 stoichiometry, the available 0.3129 mol Br_2 will react completely with 0.6258 mol NO (less than the 1.666 mol on hand):

$$0.3129 \text{ mol Br}_2 \times \frac{2 \text{ mol NO}}{\text{mol Br}_2} = 0.6258 \text{ mol NO}$$

We obtain 68.78 g NOBr in a 2:1 stoichiometric reaction:

$$50.00 \text{ g Br}_2 \times \frac{1 \text{ mol Br}_2}{159.808 \text{ g Br}_2} \times \frac{2 \text{ mol NOBr}}{\text{mol Br}_2} \times \frac{109.910 \text{ g NOBr}}{\text{mol NOBr}} = 68.78 \text{ g NOBr}$$

(b) Br_2, the limiting reactant, is consumed completely. See above.

(c) NO is present in excess:

$$\left(50.00 \text{ g NO} \times \frac{1 \text{ mol NO}}{30.006 \text{ g NO}}\right) - \left(50.00 \text{ g Br}_2 \times \frac{1 \text{ mol Br}_2}{159.808 \text{ g Br}_2} \times \frac{2 \text{ mol NO}}{\text{mol Br}_2}\right)$$

$$= 1.0406 \text{ mol NO} \quad \text{(excess)}$$

The corresponding mass, limited to four significant figures, is 31.22 g:

$$1.0406 \text{ mol NO} \times \frac{30.006 \text{ g NO}}{\text{mol NO}} = 31.22 \text{ g NO}$$

Chapter 3

Prototypical Reactions

S3-1. Identify the Brønsted-Lowry acid and base in each pair:

(a) OH^-, H_2O **(b)** H_2SO_4, HSO_4^- **(c)** NH_3, NH_4^+ **(d)** O^{2-}, OH^-

S3-2. Write a neutralization reaction in aqueous solution from which each of the following salts may arise:

(a) K_2SO_4 **(b)** $LiNO_3$ **(c)** NaBr **(d)** $BaCl_2$

S3-3. Which statement best describes the properties of water?

(a) H_2O is a Brønsted-Lowry acid.
(b) H_2O is a Brønsted-Lowry base.
(c) H_2O is neither a Brønsted-Lowry acid nor a Brønsted-Lowry base.
(d) H_2O is both a Brønsted-Lowry acid and a Brønsted-Lowry base.

Write the appropriate chemical equations to justify your choice.

S3-4. Which statement best describes the properties of the oxide ion, O^{2-}?

(a) O^{2-} is a Brønsted-Lowry acid.
(b) O^{2-} is a Brønsted-Lowry base.
(c) O^{2-} is neither a Brønsted-Lowry acid nor a Brønsted-Lowry base.
(d) O^{2-} is both a Brønsted-Lowry acid and a Brønsted-Lowry base.

Write the appropriate chemical equations to justify your choice.

S3-5. Which statement best describes the properties of the hydronium ion?

(a) H_3O^+ is a Brønsted-Lowry acid.

(b) H_3O^+ is a Brønsted-Lowry base.

(c) H_3O^+ is neither a Brønsted-Lowry acid nor a Brønsted-Lowry base.

(d) H_3O^+ is both a Brønsted-Lowry acid and a Brønsted-Lowry base.

Write the appropriate chemical equations to justify your choice.

S3-6. **(a)** Is the hydride anion, H^-, a Lewis acid or a Lewis base? Is it both? Is it neither? **(b)** Is the hydrogen cation, H^+, a Lewis acid or a Lewis base? Is it both? Is it neither?

S3-7. **(a)** Is the chloride anion, Cl^-, a Lewis acid or a Lewis base? Is it both? Is it neither? **(b)** Is the sodium cation, Na^+, a Lewis acid or a Lewis base? Is it both? Is it neither?

S3-8. In which of the following reactions does redox take place?

(a) $N_2(g) + O_2(g) \rightarrow 2NO(g)$

(b) $Ag^+(aq) + Cl^-(aq) \rightarrow AgCl(s)$

(c) $2Ag(s) + Zn^{2+}(aq) + 2OH^-(aq) \rightarrow Ag_2O(s) + Zn(s) + H_2O(\ell)$

(d) $NH_3(aq) + H_2O(\ell) \rightarrow NH_4^+(aq) + OH^-(aq)$

S3-9. Determine the oxidation number of the transition-metal atom in each species:

(a) $K_4[Fe(CN)_6]$ **(b)** $Na_2[MoOCl_4]$ **(c)** $[Pt(NH_3)_2BrCl]$ **(d)** $[CoF_6]^{3-}$

S3-10. Complete and balance the following equations:

(a) $F_2 \rightarrow F^-$

(b) $Cu^{2+} \rightarrow Cu$

(c) $Cu^{2+} \rightarrow Cu^{3+}$

(d) $OH + e^- \rightarrow$

Identify each reaction as an oxidation or reduction process. How many moles of electrons are lost or gained?

S3-11. How many moles of cations and anions will be produced when one mole of each of the following compounds is dissolved in water?

(a) Na_2SO_4 **(b)** $Ce_2(SO_4)_3$ **(c)** CsCN **(d)** LiI

Identify the cation and anion in each case.

S3-12. Each of the following formulas is incorrect:

(a) Na_2Cl **(b)** NaK **(c)** $NaBr_2$ **(d)** $LiSO_4$

Explain why.

S3-13. What mass of $LiNO_3$ is needed to produce 375 mL of a 1.07 *M* aqueous solution?

S3-14. A volume of 32.8 mL is drawn from a 1.76 *M* solution of Na_2SO_4. How many individual sodium cations and sulfate anions are contained in the 32.8 mL?

S3-15. Into what total volume of water must 4.36 g KBr be dissolved to produce a 0.578 *M* solution?

S3-16. A certain volume of 1.00 *M* $AgNO_3$ is mixed with 50.0 mL of 1.00 *M* NaCl, resulting in the precipitation of silver chloride. How many milliliters of silver nitrate must be added to produce 1.00 g AgCl? Note that NaCl is present in excess.

S3-17. The volume of a 3.00 *M* solution is doubled. What happens to the concentration?

S3-18. A 2.00 *M* solution loses one-quarter of its solvent owing to evaporation. What happens to the concentration?

S3-19. Suppose that you have 100 mL of a 0.100 *M* solution of K_2SO_4, from which you must prepare 10 mL of 0.0500 *M* K_2SO_4. How would you do so?

S3-20. Identify the free radical (if any) in each pair:

(a) N, N_2 **(b)** Ar, F **(c)** H, H^+ **(d)** H, H^-

S3-21. Identify the free radical (if any) in each pair:

(a) Cl, Cl^- **(b)** Na, Na^+ **(c)** O^-, O^{2-} **(d)** He, Li^+

S3-22. Complete and balance each of the following reactions:

(a) $CCl_2F_2 \rightarrow CClF_2 +$ __

(b) $Cl +$ __ $\rightarrow ClO$

(c) $I_2 \rightarrow I$

Which reactants and products are free radicals? Draw Lewis structures for each.

SOLUTIONS

S3-1. In the reaction of a Brønsted-Lowry acid (HA) and a Brønsted-Lowry base (B),

$$HA + B \rightarrow A^- + BH^+$$

the acid transfers a proton (H^+) to the base.

(a) H_2O, which is able to give a proton to OH^-, is the acid in this particular pair. The hydroxide ion, OH^-, is the base:

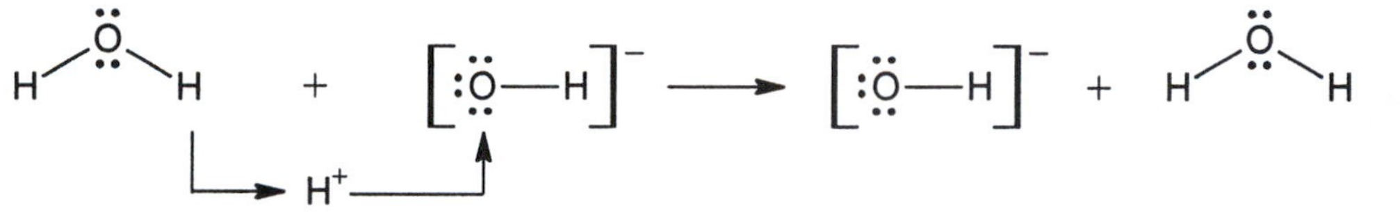

$$H_2O + [OH]^- \longrightarrow [OH]^- + H_2O \quad (H^+ \text{ transferred})$$

Viewed another way, OH^- is the *conjugate base* produced when the acid H_2O donates H^+ to some other base B:

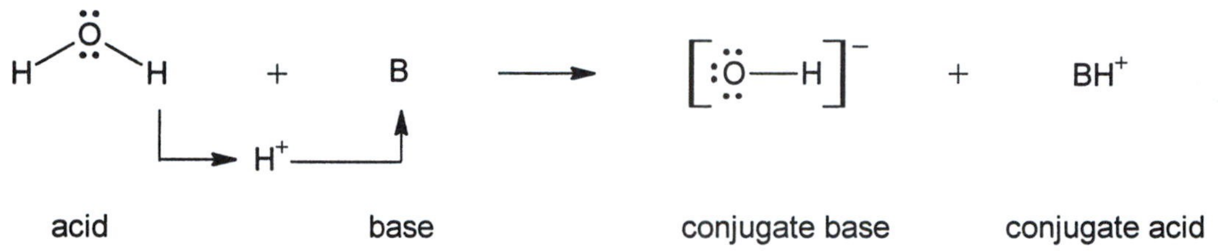

$$H_2O + B \longrightarrow [OH]^- + BH^+ \quad (H^+ \text{ transferred})$$

acid — base — conjugate base — conjugate acid

We apply similar reasoning to the three other examples in this exercise. See Section 16-2 for a treatment of conjugate acids and bases.

(b) H_2SO_4 is the acid; HSO_4^- is the base.

(c) NH_4^+ is the acid; NH_3 is the base.

(d) OH^- is the acid in this particular pair; O^{2-} is the base.

S3-2. Neutralization of an acid and base produces a salt and water.

(a) $H_2SO_4(aq) + 2KOH(aq) \rightarrow K_2SO_4(aq) + 2H_2O(\ell)$

(b) $HNO_3(aq) + LiOH(aq) \rightarrow LiNO_3(aq) + H_2O(\ell)$

(c) $HBr(aq) + NaOH(aq) \rightarrow NaBr(aq) + H_2O(\ell)$

(d) $2HCl(aq) + Ba(OH)_2(aq) \rightarrow BaCl_2(aq) + 2H_2O(\ell)$

The net ionic reaction is

$$H^+(aq) + OH^-(aq) \rightarrow H_2O(\ell)$$

or

$$H_3O^+(aq) + OH^-(aq) \rightarrow 2H_2O(\ell)$$

in each case.

S3-3. Choice (d) is correct. Water can act as both a Brønsted-Lowry acid and a Brønsted-Lowry base:

$$H_2O \text{ as an acid:} \quad H_2O + B \rightarrow OH^- + BH^+$$

$$H_2O \text{ as a base:} \quad HA + H_2O \rightarrow A^- + H_3O^+$$

S3-4. Choice (b) is correct. The oxide ion, lacking a proton to give, functions only as a Brønsted-Lowry base:

$$HA + O^{2-} \rightarrow A^- + OH^-$$

S3-5. Choice (a) is correct. The hydronium ion acts as a Brønsted-Lowry acid:

$$H_3O^+ + B \rightarrow H_2O + BH^+$$

The species H_4O^{2+} is not observed.

S3-6. In a reaction between a Lewis acid (A) and a Lewis base (B:),

$$A + B: \rightarrow A:B$$

the base donates a pair of electrons to the acid.

(a) $[H:]^-$ is an electron-pair donor, a Lewis base (and not a Lewis acid). With two electrons already filling its valence shell, the hydride ion is unable to accept an additional pair.

(b) H^+, lacking any electrons to donate, is not a Lewis base. It is a Lewis acid, able to accept a pair of electrons from a suitable partner:

$$H^+ + B: \rightarrow [H:B]^+$$

S3-7. Use the same reasoning as in the previous exercise.

(a) The chloride ion, Cl^-, is a Lewis base. It has accessible lone pairs:

$$\left[:\ddot{\underset{..}{Cl}}:\right]^-$$

Although chlorine sometimes can expand beyond an octet, the ion Cl^- generally does not accept another pair of electrons. It is not a Lewis acid.

(b) Na^+ is an electron-pair acceptor, a Lewis acid. Its valence shell stripped of electrons, the sodium ion cannot function as a Lewis base.

S3-8. Look for changes in oxidation number.

(a) The oxidation number of nitrogen increases from 0 in N_2 to +2 in NO, while oxygen goes simultaneously from 0 in O_2 to −2 in NO:

$$N_2(g) + O_2(g) \rightarrow 2NO(g)$$

Redox takes place.

(b) Redox does not occur:

$$Ag^+(aq) + Cl^-(aq) \rightarrow AgCl(s)$$

The oxidation number of silver is +1 in both $Ag^+(aq)$ and AgCl(s). The oxidation number of chlorine is unchanged as well: −1 in $Cl^-(aq)$ and AgCl(s).

(c) Silver goes from Ag^0 in its metallic form to Ag^+ in the compound Ag_2O. Zinc goes from Zn^{2+} in its cationic form to Zn^0 in the pure metal. Redox takes place:

$$2Ag(s) + Zn^{2+}(aq) + 2OH^-(aq) \rightarrow Ag_2O(s) + Zn(s) + H_2O(\ell)$$

(d) The oxidation numbers of nitrogen (–3), hydrogen (+1), and oxygen (–2) are the same in products and reactants. Redox does not take place:

$$NH_3(aq) + H_2O(\ell) \rightarrow NH_4^+(aq) + OH^-(aq)$$

Instead, a proton is transferred from H_2O to NH_3.

S3-9. The sum of the oxidation numbers of metal plus ligands must be equal to the net charge of the complex. A neutral molecule (such as NH_3) is assigned an oxidation number of 0. An ionic ligand (such as CN^-) is assigned an oxidation number equal to its net charge.

(a) Six CN^- ligands are coordinated around Fe^{2+} in the complex ion $[Fe(CN)_6]^{4-}$. Four K^+ counterions balance the total charge in the neutral coordination compound $K_4[Fe(CN)_6]$:

$$Fe^{2+} + 6CN^- \rightarrow [Fe(CN)_6]^{4-}$$

$$4K^+ + [Fe(CN)_6]^{4-} \rightarrow K_4[Fe(CN)_6]$$

Oxidation state of metal: Fe^{2+}
Net charge of complex: –4

(b) One doubly charged O^{2-} anion and four singly charged Cl^- anions are coordinated around Mo^{4+} in the complex ion $[MoOCl_4]^{2-}$. Two Na^+ counterions lie outside the coordination sphere, producing the neutral coordination compound $Na_2[MoOCl_4]$:

$$Mo^{4+} + O^{2-} + 4Cl^- \rightarrow [MoOCl_4]^{2-}$$

$$2Na^+ + [MoOCl_4]^{2-} \rightarrow Na_2[MoOCl_4]$$

Oxidation state of metal: Mo^{4+}
Net charge of complex: –2

(c) The Pt^{2+} ion, bonded to two singly charged anions (Br^-, Cl^-) and two uncharged molecules (NH_3), forms a neutral complex:

$$Pt^{2+} + 2NH_3 + Br^- + Cl^- \rightarrow [Pt(NH_3)_2BrCl]$$

Oxidation state of metal: Pt^{2+}
Net charge of complex: 0

(d) Bonded to six singly charged anions as Co^{3+}, the cobalt ion anchors a complex with an overall charge of –3:

$$Co^{3+} + 6F^- \rightarrow [CoF_6]^{3-}$$

Oxidation state of metal: Co^{3+}
Net charge of complex: –3

S3-10. Balance these simple equations first for mass and then for charge.

(a) Molecular fluorine is reduced to the fluoride ion, with two moles of electrons gained per mole of F_2:

$F_2 \rightarrow F^-$	(unbalanced)
$F_2 \rightarrow \mathbf{2}F^-$	(F is balanced)
$\mathbf{2e^-} + F_2 \rightarrow 2F^-$	(charge is balanced)

(b) One mole of Cu^{2+}, gaining two moles of electrons, is reduced to one mole of metallic copper:

$Cu^{2+} \rightarrow Cu$	(Cu is balanced)
$\mathbf{2e^-} + Cu^{2+} \rightarrow Cu$	(charge is balanced)

(c) Losing an electron, the Cu^{2+} ion is oxidized to the Cu^{3+} ion:

$Cu^{2+} \rightarrow Cu^{3+}$	(Cu is balanced)
$Cu^{2+} \rightarrow Cu^{3+} + \mathbf{e^-}$	(charge is balanced)

One mole of electrons is lost per mole of Cu^{2+}.

(d) The OH radical gains an electron and is reduced to the hydroxide ion, OH^-:

$OH + e^- \rightarrow$ __	(unbalanced)
$OH + e^- \rightarrow \mathbf{OH^-}$	(mass and charge are balanced)

One mole of electrons is gained per mole of OH.

S3-11. The exercise requires that we recognize the species SO_4^{2-} and CN^- as molecular ions.

(a) One mole of sodium sulfate dissociates into two moles of cations (Na^+) and one mole of anions (SO_4^{2-}):

$$Na_2SO_4(s) \rightarrow 2Na^+(aq) + SO_4^{2-}(aq)$$

(b) One mole of cerium(III) sulfate dissociates into two moles of Ce^{3+} cations and three moles of SO_4^{2-} anions:

$$Ce_2(SO_4)_3(s) \rightarrow 2Ce^{3+}(aq) + 3SO_4^{2-}(aq)$$

(c) Cesium cyanide releases the cesium cation and cyanide anion in 1:1 stoichiometry:

$$CsCN(s) \rightarrow Cs^+(aq) + CN^-(aq)$$

(d) One mole of lithium iodide yields one mole of lithium cations and one mole of iodide anions:

$$LiI(s) \rightarrow Li^+(aq) + I^-(aq)$$

S3-12. Anions and cations must combine in such proportions as to produce an electrically neutral compound.

(a) Sodium is a Group I metal and forms the cation Na^+. Chlorine is a Group VII nonmetal and forms the anion Cl^-. Sodium and chlorine combine in a 1:1 ratio to produce NaCl, not Na_2Cl.

(b) Sodium and potassium are both alkali metals (Group I). They do not combine with each other.

(c) Similar to (a). The sodium cation (Na^+) and bromine anion (Br^-), each singly charged, react 1:1 to produce NaBr (not $NaBr_2$).

(d) The sulfate anion, SO_4^{2-}, is doubly charged and thus requires two Li^+ cations to yield a neutral compound: Li_2SO_4.

S3-13. Use the molar mass of $LiNO_3$ to convert moles into grams:

CONCENTRATION VOLUME AMOUNT

$$\left(\frac{1.07 \text{ mol LiNO}_3}{\text{L}} \times \frac{68.946 \text{ g LiNO}_3}{\text{mol LiNO}_3}\right) \times \left(375 \text{ mL} \times \frac{1 \text{ L}}{1000 \text{ mL}}\right) = 27.7 \text{ g LiNO}_3$$

S3-14. Avogadro's number enables us to transform moles into particles:

$$\left(\frac{1.76 \text{ mol Na}_2\text{SO}_4}{1000 \text{ mL}} \times 32.8 \text{ mL}\right) \times \frac{2 \text{ mol Na}^+}{\text{mol Na}_2\text{SO}_4} \times \frac{6.022 \times 10^{23}}{\text{mol}} = 6.95 \times 10^{22} \text{ Na}^+ \text{ cations}$$

$$\left(\frac{1.76 \text{ mol Na}_2\text{SO}_4}{1000 \text{ mL}} \times 32.8 \text{ mL}\right) \times \frac{1 \text{ mol SO}_4^{2-}}{\text{mol Na}_2\text{SO}_4} \times \frac{6.022 \times 10^{23}}{\text{mol}} = 3.48 \times 10^{22} \text{ SO}_4^{2-} \text{ anions}$$

S3-15. Another straightforward application of the unit-factor method, mediated by the molar mass:

$$\frac{1000 \text{ mL}}{0.578 \text{ mol KBr}} \times \frac{1 \text{ mol KBr}}{119.002 \text{ g KBr}} \times 4.36 \text{ g KBr} = 63.4 \text{ mL}$$

S3-16. All parties to the precipitation interact in 1:1 stoichiometry:

$$AgNO_3(aq) + NaCl(aq) \rightarrow AgCl(s) + NaNO_3(aq)$$

Stringing together the unit factors, we convert grams of AgCl into milliliters of $AgNO_3$:

$$1.00 \text{ g AgCl} \times \frac{1 \text{ mol AgCl}}{143.321 \text{ g AgCl}} \times \frac{1 \text{ mol AgNO}_3}{\text{mol AgCl}} \times \frac{1000 \text{ mL AgNO}_3}{1.00 \text{ mol AgNO}_3} = 6.98 \text{ mL AgNO}_3$$

S3-17. Suppose that we have 1.00 L. Since the amount of dissolved material (3.00 mol) remains the same, the concentration is halved when the volume is doubled:

$$\frac{3.00 \text{ mol}}{1.00 \text{ L}} \rightarrow \frac{3.00 \text{ mol}}{2.00 \text{ L}} = 1.50\ M$$

Note that these relationships, in general, are governed by the following equation:

$$c_1V_1 = c_2V_2$$

initial concentration (c_1), initial volume (V_1), final concentration (c_2), final volume (V_2)

The product of concentration and volume has dimensions of *amount*:

Concentration × volume = amount

$$\frac{\text{mol}}{\text{L}} \times \text{L} = \text{mol}$$

S3-18. Similar. The solution becomes 33% more concentrated when one-quarter of the solvent evaporates:

$$\frac{2.00 \text{ mol}}{1.00 \text{ L}} \rightarrow \frac{2.00 \text{ mol}}{0.75 \text{ L}} = 2.67\ M$$

Equivalently:

$$c_1V_1 = c_2V_2$$

$$c_2 = \frac{c_1V_1}{V_2} = \frac{\left(2.00 \text{ mol L}^{-1}\right)V_1}{0.75\, V_1} = 2.67\ M$$

S3-19. Dilute 5 mL of the original 0.100 *M* solution to a total volume of 10 mL. With the molar amount of K_2SO_4 fixed but the volume doubled, the concentration is cut in half (to 0.0500 *M*).

Various other approaches are also possible. For example, we can dilute the 0.100 *M* solution from 100 mL to 200 mL and then withdraw 10 mL. The concentration of K_2SO_4 is uniformly 0.0500 *M* in both the 200-mL and 10-mL portions.

S3-20. A free radical accommodates one or more unpaired electrons in its valence shell.

(a) An atom of nitrogen, N, has an odd number of electrons and is thus a free radical. Its open shell contains three unpaired electrons (as will be explained in Chapter 6):

·Ṅ·
 ··

A molecule of nitrogen, N_2, contains no unpaired electrons. Ten valence electrons are distributed as three bonding pairs and two lone pairs:

:N≡N:

N_2 is a closed-shell molecule, not a free radical.

(b) Argon, a noble gas, contains no unpaired electrons in its completed octet. It is not a free radical:

:Ar:

An atom of fluorine, however, has an open shell with one unpaired electron:

:F·

The F atom is a free radical.

(c) The hydrogen atom, H·, contains a single electron (by definition, unpaired) and is clearly a free radical. The bare proton H^+, containing *no* electrons, is not.

(d) The hydride ion, $[H{:}]^-$, contains two electrons in a closed-shell configuration isoelectronic with He:. It is not a free radical.

The hydrogen *atom* is a free radical, as explained above in (c).

S3-21. More radicals.

(a) The chlorine atom remains one electron short of an octet and is therefore a free radical:

$$:\ddot{\underset{\cdot\cdot}{\mathrm{Cl}}}\cdot$$

The chloride ion, Cl^-, is isoelectronic with Ar and contains no unpaired electrons in its completed octet:

$$\left[:\ddot{\underset{\cdot\cdot}{\mathrm{Cl}}}:\right]^-$$

Cl^- is not a free radical.

(b) The sodium atom, Na, contains just one valence electron and consequently is a free radical. The sodium ion, Na^+, is isoelectronic with Ne. It has no unpaired electrons and is not a free radical.

(c) The anion O^-, one electron shy of an octet, is a free radical isoelectronic with F:

$$\left[:\ddot{\underset{\cdot\cdot}{\mathrm{O}}}\cdot\right]^-$$

The oxide ion, O^{2-}, has a completed octet and is isoelectronic with Ne:

$$\left[:\ddot{\underset{\cdot\cdot}{\mathrm{O}}}:\right]^{2-}$$

It is not a free radical.

(d) The lithium ion, Li^+, is isoelectronic with He. Both species contain two paired electrons in a closed shell. Neither is a radical.

S3-22. One more exercise dealing with free radicals.

(a) CCl_2F_2 dissociates into $CClF_2$ and Cl. Both products are radicals:

$$\mathrm{Cl{-}CF_2{-}Cl} \longrightarrow \mathrm{Cl{-}\dot{C}F_2} + \mathrm{Cl\cdot}$$

(b) Cl and O combine to produce ClO. All three species are free radicals:

$$:\ddot{\underset{\cdot\cdot}{\mathrm{Cl}}}\cdot + \cdot\ddot{\underset{\cdot\cdot}{\mathrm{O}}}\cdot \longrightarrow :\ddot{\underset{\cdot\cdot}{\mathrm{Cl}}}{-}\ddot{\underset{\cdot\cdot}{\mathrm{O}}}\cdot$$

ClO may also be represented by the equivalent Lewis structure shown below:

$$\cdot\ddot{\underset{\cdot\cdot}{\mathrm{Cl}}}{-}\ddot{\underset{\cdot\cdot}{\mathrm{O}}}:$$

The electron configuration of oxygen, a diradical, is discussed in Chapter 6.

(c) The iodine molecule dissociates into two iodine atoms:

$$I_2 \rightarrow 2I$$

The species I, with seven valence electrons, is a free radical:

Chapter 4

Light and Matter—Waves and/or Particles

S4-1. The speed of light in vacuum is 670.6 million miles per hour. Calculate the frequency (in s^{-1}) corresponding to each of the following electromagnetic wavelengths:

(a) 1.00 mi **(b)** 1.00 ft **(c)** 1.00 in

S4-2. One *light-year* is defined as the distance traversed in one year by electromagnetic radiation traveling in vacuum. What is its value in **(a)** meters, **(b)** kilometers, **(c)** centimeters, **(d)** inches, **(e)** feet, and **(f)** miles?

S4-3. **(a)** What distance does a beam of light traveling in vacuum advance in 1.000 s? **(b)** Suppose that the beam bounces back and forth between two mirrors separated by 0.500 m. How much time is required for one round trip?

S4-4. Choose an arbitrary wavelength and frequency for a traveling wave. **(a)** Sketch three cycles of the oscillation, indicating the extent of the wavelength on the diagram. **(b)** Sketch three cycles for a wave traveling at the same speed but half the original frequency. **(c)** Sketch three cycles for a wave traveling at the same speed but twice the original frequency.

S4-5. Suppose that the frequency of one electromagnetic wave is ν_1 and the frequency of another electromagnetic wave traveling at the same speed is ν_2. What is the ratio of their wavelengths?

S4-6. Say that two pulses of red light originate from two different sources but have identical wavelengths ($\lambda = 700$ nm). They meet at the same point and at the same time, with one pulse having traveled a distance L_1 and the other having traveled a distance L_2. **(a)** If the signals arrive exactly 180° out of phase, what is the minimum value of the difference $|L_2 - L_1|$? **(b)** Repeat the calculation for a phase difference of 90°. **(c)** Do it again for 270°.

S4-7. A source produces X rays with a frequency of 3.00×10^{18} Hz. **(a)** What is the period of the wave (the time per cycle)? **(b)** How much of a time lag will produce a phase difference of 30° between two such waves?

S4-8. **(a)** Use the equation $E = h\nu$ to express energy in units of kg, m, and s. **(b)** Use the equation $p = h/\lambda$ to express momentum in units of kg, m, and s. **(c)** Show that the combination pc has units of energy. **(d)** Show that the combination mc has units of momentum.

S4-9. Is it possible to have an equation in which a quantity with dimensions *energy* × *time* is set equal to a quantity with dimensions *momentum* × *length*?

S4-10. Define a quantity $\bar{\nu}$ called the *wavenumber*:

$$\bar{\nu} = \frac{1}{\lambda}$$

(a) Show that the wavenumber is directly proportional to photon energy:

$$E = hc\bar{\nu}$$

(b) Compute the value of $\bar{\nu}$, in units of cm^{-1}, for a photon with frequency equal to 3.00×10^{14} Hz.

S4-11. **(a)** Compute the energy of a photon with a wavenumber of 1000 cm^{-1}. **(b)** Do the same for a wavenumber of 500 cm^{-1}. **(c)** Which of the two photons has the higher frequency? **(d)** Which of the two photons has the longer wavelength?

S4-12. Which photons carry more energy?

(a) X ray or gamma ray
(b) red or green
(c) microwave or infrared
(d) violet or ultraviolet

S4-13. Which photons carry more momentum?

(a) blue or yellow
(b) infrared or ultraviolet
(c) radio or microwave
(d) red or infrared

S4-14. **(a)** A pulse of light delivers 1.00 J of electromagnetic energy at a wavelength of 1987 Å. Approximately how many photons are contained in the pulse? **(b)** Approximately how many photons are contained in a pulse of the same frequency that delivers only 1.00×10^{-16} J?

S4-15. Chromium has a work function of 7.21×10^{-19} J. Calculate the maximum wavelength able to expel a photoelectron.

S4-16. Illuminated at a wavelength of 4672 Å, a surface of europium emits a photoelectron with kinetic energy equal to 2.47×10^{-20} J. Calculate the work function of europium.

S4-17. The work function of cobalt is 8.01×10^{-19} J. Calculate the electromagnetic wavelength that will impart a velocity of 2.04×10^{5} m s^{-1} to an outgoing photoelectron.

S4-18. Consider a one-dimensional standing wave confined over a distance of 1 meter. **(a)** Sketch the third harmonic. **(b)** How many nodes are present? Specify, in meters, the location of each node.

S4-19. Rank the following particles in order of increasing de Broglie wavelength, assuming each to be moving at the same velocity: proton, electron, neutron, deuterium nucleus, methane molecule (CH_4).

S4-20. Assume that an electron is moving at 1.00×10^{6} m s^{-1}. At what velocity will a neutron have the same de Broglie wavelength? Note that the mass of a neutron is 1.675×10^{-27} kg.

S4-21. An alternative form of the indeterminacy principle expresses the relationship between uncertainty in energy and uncertainty in time:

$$\Delta E \, \Delta t \geq \frac{h}{4\pi}$$

Suppose that a particular quantum state has a lifetime of 1.0×10^{-8} s. Estimate the corresponding uncertainty in its energy.

SOLUTIONS

S4-1. Given the wavelength λ and speed c, we calculate the corresponding frequency ν in any consistent set of units:

$$\lambda\nu = c$$

(a) The only conversion necessary is from hours into seconds:

$$\nu = \frac{c}{\lambda} = \frac{\left(\dfrac{670.6 \times 10^6 \text{ mi}}{\text{h}} \times \dfrac{1 \text{ h}}{3600 \text{ s}}\right)}{1.00 \text{ mi}} = 1.86 \times 10^5 \text{ s}^{-1}$$

(b) One mile is equal to exactly 5280 ft:

$$\nu = \frac{c}{\lambda} = \frac{\left(\dfrac{670.6 \times 10^6 \text{ mi}}{\text{h}} \times \dfrac{1 \text{ h}}{3600 \text{ s}}\right)}{\left(1.00 \text{ ft} \times \dfrac{1 \text{ mi}}{5280 \text{ ft}}\right)} = 9.84 \times 10^8 \text{ s}^{-1}$$

(c) Similar. One foot is equal to exactly 12 in:

$$\nu = \frac{c}{\lambda} = \frac{\left(\dfrac{670.6 \times 10^6 \text{ mi}}{\text{h}} \times \dfrac{1 \text{ h}}{3600 \text{ s}}\right)}{\left(1.00 \text{ in} \times \dfrac{1 \text{ ft}}{12 \text{ in}} \times \dfrac{1 \text{ mi}}{5280 \text{ ft}}\right)} = 1.18 \times 10^{10} \text{ s}^{-1}$$

S4-2. More practice with unit conversions. We begin, in each case, with the speed of light in vacuum,

$$c = 2.998 \times 10^8 \text{ m s}^{-1}$$

and define one year as 365.25 days.

(a) $$\frac{2.998 \times 10^8 \text{ m}}{\text{s}} \times \frac{86{,}400 \text{ s}}{\text{d}} \times \frac{365.25 \text{ d}}{\text{y}} \times 1 \text{ y} = 9.46 \times 10^{15} \text{ m}$$

(b) $$\frac{2.998 \times 10^8 \text{ m}}{\text{s}} \times \frac{1 \text{ km}}{1000 \text{ m}} \times \frac{86{,}400 \text{ s}}{\text{d}} \times \frac{365.25 \text{ d}}{\text{y}} \times 1 \text{ y} = 9.46 \times 10^{12} \text{ km}$$

(c) $$\frac{2.998 \times 10^8 \text{ m}}{\text{s}} \times \frac{100 \text{ cm}}{\text{m}} \times \frac{86{,}400 \text{ s}}{\text{d}} \times \frac{365.25 \text{ d}}{\text{y}} \times 1 \text{ y} = 9.46 \times 10^{17} \text{ cm}$$

(d) $$\frac{2.998\times10^{8}\ \text{m}}{\text{s}}\times\frac{100\ \text{cm}}{\text{m}}\times\frac{1\ \text{in}}{2.54\ \text{cm}}\times\frac{86{,}400\ \text{s}}{\text{d}}\times\frac{365.25\ \text{d}}{\text{y}}\times 1\ \text{y} = 3.72\times10^{17}\ \text{in}$$

(e) $$\frac{2.998\times10^{8}\ \text{m}}{\text{s}}\times\frac{100\ \text{cm}}{\text{m}}\times\frac{1\ \text{in}}{2.54\ \text{cm}}\times\frac{1\ \text{ft}}{12\ \text{in}}\times\frac{86{,}400\ \text{s}}{\text{d}}\times\frac{365.25\ \text{d}}{\text{y}}\times 1\ \text{y}$$

$$= 3.10\times10^{16}\ \text{ft}$$

(f) $$\frac{2.998\times10^{8}\ \text{m}}{\text{s}}\times\frac{100\ \text{cm}}{\text{m}}\times\frac{1\ \text{in}}{2.54\ \text{cm}}\times\frac{1\ \text{ft}}{12\ \text{in}}\times\frac{1\ \text{mi}}{5280\ \text{ft}}\times\frac{86{,}400\ \text{s}}{\text{d}}\times\frac{365.25\ \text{d}}{\text{y}}\times 1\ \text{y}$$

$$= 5.88\times10^{12}\ \text{mi}$$

S4-3. The speed of light in vacuum is 2.998×10^{8} m s^{-1}, independent of frequency.

(a) $$\frac{2.998\times10^{8}\ \text{m}}{\text{s}}\times 1.000\ \text{s} = 2.998\times10^{8}\ \text{m}$$

(b) $$\frac{1\ \text{s}}{2.998\times10^{8}\ \text{m}}\times\left(2\times 0.500\ \text{m}\right) = 3.34\times10^{-9}\ \text{s}$$

S4-4. The product of wavelength and frequency is equal to a constant, the speed of light:

$$\lambda\nu = c$$

(a) The wavelength is the distance between successive crests or troughs:

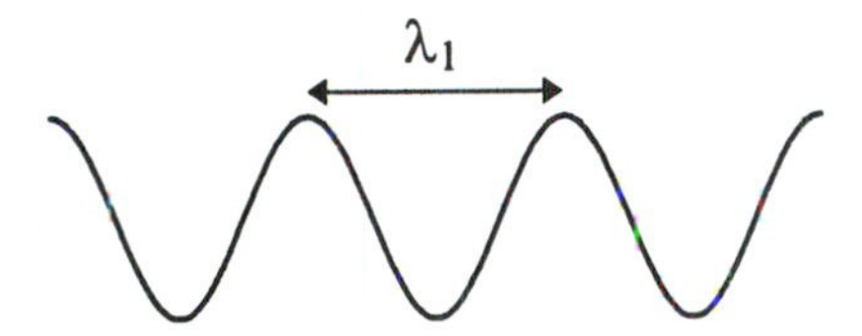

(b) If the frequency is halved, then the wavelength is doubled:

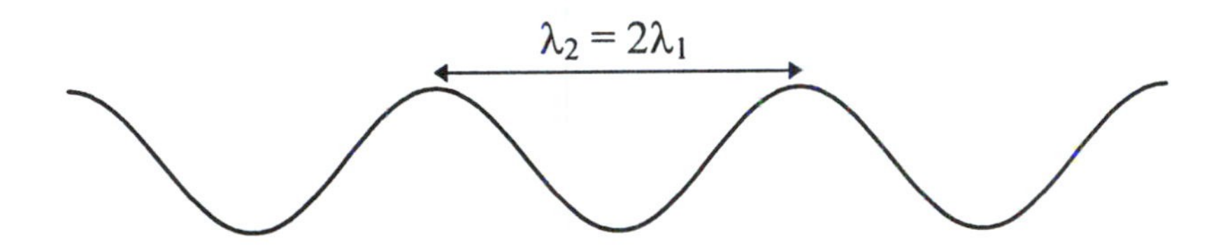

(c) If the frequency is doubled, then the wavelength is halved:

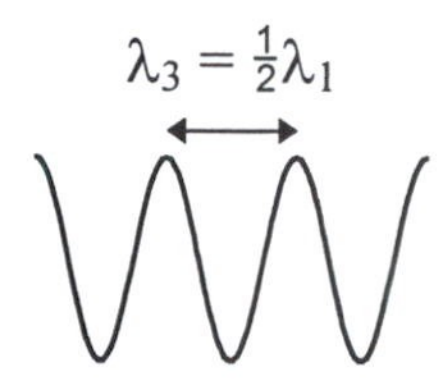

S4-5. Wavelength and frequency are inversely proportional:

$$\lambda_1 \nu_1 = \lambda_2 \nu_2$$

$$\frac{\lambda_1}{\lambda_2} = \frac{\nu_2}{\nu_1}$$

S4-6. An oscillation undergoes a phase change of 360° (2π radians) over the course of one complete cycle.

(a) A phase difference of 180° corresponds to $\frac{1}{2}$ cycle:

$$|L_1 - L_2| = \frac{\lambda}{2} = \frac{700 \text{ nm}}{2} = 350 \text{ nm}$$

(b) A phase difference of 90° corresponds to $\frac{1}{4}$ cycle:

$$|L_1 - L_2| = \frac{\lambda}{4} = \frac{700 \text{ nm}}{4} = 175 \text{ nm}$$

(c) A phase difference of 270° corresponds to $\frac{3}{4}$ cycle:

$$|L_1 - L_2| = \frac{3}{4}\lambda = \frac{3}{4} \times 700 \text{ nm} = 525 \text{ nm}$$

S4-7. The period of a wave, T, is the reciprocal of its frequency:

$$T = \frac{1}{\nu}$$

(a) Given a frequency, we calculate the period:

$$T = \frac{1}{\nu} = \frac{1}{3.00 \times 10^{18} \text{ s}^{-1}} = 3.33 \times 10^{-19} \text{ s}$$

(b) A phase difference of 30° corresponds to $\frac{1}{12}$ cycle:

$$\frac{T}{12} = \frac{3.33 \times 10^{-19}\ \text{s}}{12} = 2.78 \times 10^{-20}\ \text{s}$$

S4-8. Force (measured in newtons) is defined by Newton's second law,

$$F = ma \sim \text{kg m s}^{-2} \sim \text{N}$$

and work/energy (measured in joules) is defined as the product of force and distance:

$$W = Fd \sim (\text{kg m s}^{-2})\ \text{m} \sim \text{kg m}^2\ \text{s}^{-2} \sim \text{J}$$

Momentum, p, is the product of mass and velocity:

$$p = mv \sim \text{kg m s}^{-1}$$

(a) $E = h\nu \sim (\text{J s})(\text{s}^{-1}) \sim \text{J} \sim \text{kg m}^2\ \text{s}^{-2}$

(b) $p = \dfrac{h}{\lambda} \sim \dfrac{\text{J s}}{\text{m}} \sim \dfrac{\left(\text{kg m}^2\ \text{s}^{-2}\right)\text{s}}{\text{m}} \sim \text{kg m s}^{-1}$

(c) $pc \sim (\text{kg m s}^{-1})(\text{m s}^{-1}) \sim \text{kg m}^2\ \text{s}^{-2} \sim \text{J}$

(d) $mc \sim \text{kg m s}^{-1}$

S4-9. Yes, such a combination is legitimate. The quantities *energy* × *time* and *momentum* × *length* have the same units:

$$\text{Energy} \times \text{time} \sim \text{J s} \sim (\text{kg m}^2\ \text{s}^{-2})\ \text{s} \sim \text{kg m}^2\ \text{s}^{-1}$$

$$\text{Momentum} \times \text{length} \sim (\text{kg m s}^{-1})\ \text{m} \sim \text{kg m}^2\ \text{s}^{-1}$$

S4-10. We define a wavenumber $\bar{\nu}$ (the number of cycles per unit length) as the reciprocal of wavelength:

$$\bar{\nu} = \frac{1}{\lambda}$$

(a) Combine $\lambda\nu = c$ with the Planck-Einstein relationship

$$E = h\nu$$

to show that energy is directly proportional to wavenumber:

$$E = h\nu = \frac{hc}{\lambda} = hc\bar{\nu}$$

(b) Given the frequency, we calculate the wavelength and associated wavenumber:

$$\bar{\nu} = \frac{1}{\lambda} = \frac{1}{(c/\nu)} = \frac{\nu}{c} = \frac{3.00 \times 10^{14}\ \text{s}^{-1}}{2.998 \times 10^{10}\ \text{cm s}^{-1}} = 1.00 \times 10^{4}\ \text{cm}^{-1}$$

Note that the speed of light is conveniently expressed in cm s^{-1}.

S4-11. We know from the preceding exercise that photon energy is proportional to wavenumber:

$$E = hc\bar{\nu}$$

(a) Expressing the speed of light in cm s^{-1}, we maintain a consistent set of units:

$$E = hc\bar{\nu} = (6.626 \times 10^{-34}\ \text{J s})(2.998 \times 10^{10}\ \text{cm s}^{-1})(1000\ \text{cm}^{-1}) = 1.986 \times 10^{-20}\ \text{J}$$

(b) Similar:

$$E = hc\bar{\nu} = (6.626 \times 10^{-34}\ \text{J s})(2.998 \times 10^{10}\ \text{cm s}^{-1})(500\ \text{cm}^{-1}) = 9.93 \times 10^{-21}\ \text{J}$$

(c) The photon with the higher wavenumber (1000 cm^{-1}) has the higher energy and higher frequency.

(d) The photon with the lower wavenumber (500 cm^{-1}) has the lower energy and longer wavelength.

S4-12. The electromagnetic spectrum is ordered in the following way:

$$E(\text{radio}) < E(\mu) < E(\text{ir}) < E(\text{vis}) < E(\text{uv}) < E(\text{X}) < E(\gamma)$$

Within the visible portion, energies increase from red to violet:

$$E(\text{R}) < E(\text{O}) < E(\text{Y}) < E(\text{G}) < E(\text{B}) < E(\text{V})$$

(a) $E(\gamma) > E(\mathrm{X})$

(b) $E(\text{vis-G}) > E(\text{vis-R})$

(c) $E(\text{ir}) > E(\mu)$

(d) $E(\text{uv}) > E(\text{vis-V})$

S4-13. Momentum is inversely proportional to wavelength,

$$p = \frac{h}{\lambda}$$

and therefore directly proportional to frequency and energy:

$$p = \frac{h}{\lambda} = \frac{h\nu}{c} = \frac{E}{c}$$

See Exercise S4-12 for the progression of energy in the electromagnetic spectrum.

(a) $p(\text{vis-B}) > p(\text{vis-Y})$

(b) $p(\text{uv}) > p(\text{ir})$

(c) $p(\mu) > p(\text{radio})$

(d) $p(\text{vis-R}) > p(\text{ir})$

S4-14. The total energy carried by N photons, E_N, is the sum of N quanta (each equal to $h\nu$):

$$E_N = Nh\nu = \frac{Nhc}{\lambda}$$

(a) Inserting the wavelength λ (given as 1987 Å = 1.987×10^{-7} m) and the energy (1.00 J), we solve for N:

$$N = \frac{\lambda E_N}{hc} = \frac{\left(1.987 \times 10^{-7} \text{ m}\right)\left(1.00 \text{ J}\right)}{\left(6.626 \times 10^{-34} \text{ J s}\right)\left(2.998 \times 10^{8} \text{ m s}^{-1}\right)} = 1.00 \times 10^{18} \text{ photons}$$

The number is large enough to make the electromagnetic field appear quasi-continuous (unquantized).

(b) Similar, for a total energy of 1.00×10^{-16} J:

$$N = \frac{\lambda E_N}{hc} = \frac{(1.987 \times 10^{-7}\ \text{m})(1.00 \times 10^{-16}\ \text{J})}{(6.626 \times 10^{-34}\ \text{J s})(2.998 \times 10^{8}\ \text{m s}^{-1})} = 1.00 \times 10^{2}\ \text{photons}$$

At this low level of energy, the discrete nature of the electromagnetic field starts to become clear.

S4-15. Let E_0 and λ_0 denote the work function and corresponding threshold wavelength, respectively. Only photons with wavelengths less than or equal to λ_0 will be able to eject a photoelectron:

$$E_0 = h\nu_0 = \frac{hc}{\lambda_0}$$

$$\lambda_0 = \frac{hc}{E_0} = \frac{(6.626 \times 10^{-34}\ \text{J s})(2.998 \times 10^{8}\ \text{m s}^{-1})}{7.21 \times 10^{-19}\ \text{J}} = 2.76 \times 10^{-7}\ \text{m}$$

The threshold is reached at 276 nm (equivalently, 2760 Å).

S4-16. Photon energy ($E = h\nu = hc/\lambda$) in excess of the work function (E_0) is imparted to the photoelectron in the form of kinetic energy (E_k):

$$\frac{hc}{\lambda} = E_0 + E_k$$

$$E_0 = \frac{hc}{\lambda} - E_k = \frac{(6.626 \times 10^{-34}\ \text{J s})(2.998 \times 10^{8}\ \text{m s}^{-1})}{4672 \times 10^{-10}\ \text{m}} - 2.47 \times 10^{-20}\ \text{J} = 4.005 \times 10^{-19}\ \text{J}$$

Recall that 1 Å = 10^{-10} m.

S4-17. Similar to the preceding exercise, except now we substitute $E_k = \frac{1}{2}mv^2$ for the kinetic energy and solve for the photon wavelength:

$$\frac{hc}{\lambda} = E_0 + \frac{1}{2}mv^2$$

$$\lambda = \frac{hc}{E_0 + \frac{1}{2}mv^2} = \frac{(6.626 \times 10^{-34}\ \text{J s})(2.998 \times 10^{8}\ \text{m s}^{-1})}{(8.01 \times 10^{-19}\ \text{J}) + \frac{1}{2}(9.109 \times 10^{-31}\ \text{kg})(2.04 \times 10^{5}\ \text{m s}^{-1})^2}$$

$$= 2.42 \times 10^{-7}\ \text{m} \quad (242\ \text{nm})$$

S4-18. The nth harmonic of a standing wave confines n half-wavelengths within a space L. There are $n - 1$ nodes, not including the two endpoints.

(a) The first and third harmonics are shown below for a system where $L = 1$ m:

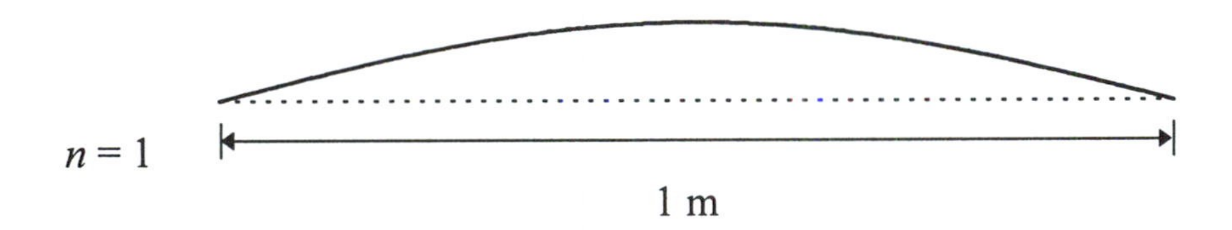

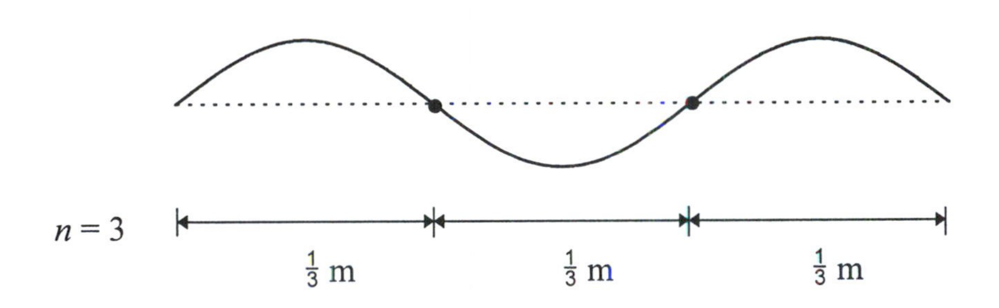

(b) The third harmonic contains two nodes, located at $\frac{1}{3}$ m and $\frac{2}{3}$ m along the string.

S4-19. The de Broglie wavelength is inversely proportional to momentum:

$$\lambda_{\text{deB}} = \frac{h}{mv}$$

For two particles moving at the same velocity, λ_{deB} is therefore inversely proportional to mass. The wavelengths increase in the following order, from the particle with the largest mass to the particle with the smallest:

$$\lambda_{\text{deB}}(\text{CH}_4) < \lambda_{\text{deB}}(^2\text{H}) < \lambda_{\text{deB}}(\text{n}) < \lambda_{\text{deB}}(\text{p}) < \lambda_{\text{deB}}(\text{e}^-)$$

A neutron (1.67493×10^{-27} kg) is slightly more massive than a proton (1.67262×10^{-27} kg).

S4-20. The more massive particle, the neutron, moves at a slower speed to acquire the same de Broglie wavelength as the electron:

$$\frac{h}{m_e v_e} = \frac{h}{m_n v_n}$$

$$v_n = \frac{m_e v_e}{m_n} = \frac{(9.109 \times 10^{-31}\ \text{kg})(1.00 \times 10^6\ \text{m s}^{-1})}{1.675 \times 10^{-27}\ \text{kg}} = 544\ \text{m s}^{-1}$$

S4-21. The uncertainties in energy and time are inversely proportional:

$$\Delta E \, \Delta t \geq \frac{h}{4\pi}$$

$$\Delta E \geq \frac{h}{4\pi \, \Delta t} = \frac{6.626 \times 10^{-34} \text{ J s}}{4\pi\left(1.0 \times 10^{-8} \text{ s}\right)} \approx 5.3 \times 10^{-27} \text{ J}$$

Chapter 5

Quantum Theory of the Hydrogen Atom

S5-1. Pick the subshell in which a hydrogen electron will have the larger orbital angular momentum:

(a) $2p$, $4s$ **(b)** $3p$, $3d$ **(c)** $1s$, $2p$ **(d)** $4f$, $8s$

S5-2. Repeat the exercise for He^+. In which subshell will the electron have the larger orbital angular momentum?

(a) $2p$, $4s$ **(b)** $3p$, $3d$ **(c)** $1s$, $2p$ **(d)** $4f$, $8s$

S5-3. Pick the subshell in which a hydrogen electron will have the larger energy:

(a) $4d$, $5s$ **(b)** $2s$, $2p$ **(c)** $3d$, $6s$ **(d)** $6s$, $6p$

S5-4. Again, for He^+: In which subshell will the electron have the larger energy?

(a) $4d$, $5s$ **(b)** $2s$, $2p$ **(c)** $3d$, $6s$ **(d)** $6s$, $6p$

S5-5. How many spatial orbitals are contained within the $n = 7$ shell of a hydrogen atom? List them.

S5-6. How many values of m_ℓ correspond to the quantum number $\ell = 5$? List them.

S5-7. How many angular nodes are present in a $7g$ orbital? How many radial nodes?

S5-8. State the values of n and ℓ that correspond to each subshell:

(a) $5g$ **(b)** $7s$ **(c)** $6p$ **(d)** $9d$

S5-9. For each subshell in the preceding exercise, list all possible values of m_ℓ:

(a) $5g$ **(b)** $7s$ **(c)** $6p$ **(d)** $9d$

S5-10. Explain why each of the following combinations of n and ℓ is forbidden:

(a) $2f$ **(b)** $4g$ **(c)** $3h$ **(d)** $1p$

S5-11. Give the name of the subshell corresponding to each combination of n and ℓ:

(a) $n = 7,\ \ell = 5$
(b) $n = 4,\ \ell = 2$
(c) $n = 6,\ \ell = 0$
(d) $n = 5,\ \ell = 1$

S5-12. Which sets of quantum numbers are forbidden?

(a) $n = -1,\ \ell = 0,\ m_\ell = 0,\ m_s = \frac{1}{2}$
(b) $n = 1,\ \ell = 1,\ m_\ell = 0,\ m_s = \frac{1}{2}$
(c) $n = 5,\ \ell = 4,\ m_\ell = 5,\ m_s = -\frac{1}{2}$
(d) $n = 5,\ \ell = 5,\ m_\ell = 4,\ m_s = -\frac{1}{2}$

For each violation, state the reasons why.

S5-13. Which sets of quantum numbers are forbidden?

(a) $n = 1,\ \ell = 0,\ m_\ell = 0,\ m_s = \frac{1}{4}$
(b) $n = 0,\ \ell = 1,\ m_\ell = 0,\ m_s = \frac{1}{2}$
(c) $n = 3,\ \ell = 2,\ m_\ell = -3,\ m_s = -\frac{1}{2}$
(d) $n = \frac{3}{2},\ \ell = \frac{1}{2},\ m_\ell = \frac{1}{2},\ m_s = \frac{1}{2}$

For each violation, state the reasons why.

S5-14. Which sets of quantum numbers are forbidden?

(a) $n = 1,\ \ell = 0,\ m_\ell = 0,\ m_s = 1$
(b) $n = 1,\ \ell = 1,\ m_\ell = 0,\ m_s = 0$
(c) $n = 3,\ \ell = 2,\ m_\ell = -2,\ m_s = -\frac{1}{2}$
(d) $n = 10{,}000,\ \ell = 1,\ m_\ell = 0,\ m_s = \frac{1}{2}$

For each violation, state the reasons why.

S5-15. The *Bohr theory*, an early model of the one-electron atom, treats the electron as a "planet" orbiting a nuclear "sun" in a quantized, electromagnetic solar system. By restricting the electronic angular momentum to the values $nh/2\pi$ (where $n = 1, 2, 3, \ldots$), Bohr argued that the electron can persist indefinitely in circular orbits of fixed radii—in clear defiance of the laws of classical mechanics and electromagnetism. The idea worked, too, but only up to a certain point: Although the quantized energies he predicted were correct, this "old quantum theory" of fixed orbits soon proved unsuitable for systems more complex than one-electron atoms. In what way is the Bohr model in conflict with our present-day understanding of quantum mechanics?

S5-16. The radius of a Bohr orbit is given by the formula below:

$$r_n = \frac{\varepsilon_0 n^2 h^2}{\pi Z e^2 m_e} = \frac{n^2}{Z} a_0$$

The symbols ε_0, h, e, and m_e denote, respectively, the permittivity of vacuum (Chapter 1), Planck's constant, the magnitude of the electronic charge, and the mass of the electron. Numerical values for all these physical constants are tabulated in Appendix C of *Principles of Chemistry*. **(a)** Compute a_0. **(b)** Compute the orbital radius of a hydrogen electron in the first shell. **(c)** Do the same for a hydrogen electron in the second shell. **(d)** Once more: for a hydrogen electron in the third shell.

S5-17. Repeat the preceding exercise for He^+: Calculate r_1, r_2, and r_3, and explain the trend.

S5-18. Do the same for Li^{2+}.

S5-19. A Bohr electron orbiting with radius r_n has the following velocity:

$$v_n = \frac{Ze^2}{2\varepsilon_0 nh}$$

Write an equation for the corresponding kinetic energy.

S5-20. The electrostatic potential energy between the electron and nucleus in a one-electron atom is given below:

$$E_{\text{pot}} = -\frac{Ze^2}{4\pi\varepsilon_0 r}$$

(a) Combine this expression with the kinetic energy of an orbiting Bohr electron (from the preceding exercise) to show that the total energy is quantized as follows:

$$E_n = -\frac{m_e Z^2 e^4}{8\varepsilon_0^2 n^2 h^2}$$

(b) Insert values of the physical constants into the expression in (a), and thus show that E_n is numerically equivalent to the consolidated formula below:

$$E_n = -\frac{Z^2}{n^2} R_\infty$$

S5-21. Consider again the energy equation,

$$E_n = -\frac{m_e Z^2 e^4}{8\varepsilon_0^2 n^2 h^2}$$

and note that the Balmer wavelengths for hydrogen are 656 nm, 486 nm, 434 nm, and 410 nm. What emission wavelengths would be observed if the mass of an electron were suddenly doubled?

S5-22. Repeat the preceding exercise, but this time pretend that the *charge* of an electron is doubled from its usual value. In what ways will the Balmer wavelengths for hydrogen be affected?

S5-23. Imagine, once more, what would happen if the physical constants m_e and e were to depart from their usual values. **(a)** Compute the ionization energies for H and He^+ that would result if the mass of the electron were doubled. Explain the relationship. **(b)** Compute the ionization energies that would result if the electronic charge were doubled.

SOLUTIONS

S5-1. The higher the value of ℓ, the larger is the orbital angular momentum:

$$\begin{array}{ll} s & \ell = 0 \\ p & \ell = 1 \\ d & \ell = 2 \\ f & \ell = 3 \end{array}$$

Measured in units of $h/2\pi$, the angular momentum has an average magnitude of $\sqrt{\ell(\ell+1)}$.

(a) $\ell(2p) > \ell(4s)$
(b) $\ell(3d) > \ell(3p)$
(c) $\ell(2p) > \ell(1s)$
(d) $\ell(4f) > \ell(8s)$

The principal quantum number, n, is irrelevant here.

S5-2. The ranking of orbital angular momentum is the same in He^+ as it is in H. Both systems contain only one electron and one nucleus.

(a) $\ell(2p) > \ell(4s)$
(b) $\ell(3d) > \ell(3p)$
(c) $\ell(2p) > \ell(1s)$
(d) $\ell(4f) > \ell(8s)$

S5-3. The energy of a one-electron atom, proportional to $-n^{-2}$, is determined solely by the principal quantum number. The higher the value of n, the greater (more positive) is the energy.

(a) $n(5s) > n(4d)$
(b) $n(2s) = n(2p)$
(c) $n(6s) > n(3d)$
(d) $n(6s) = n(6p)$

S5-4. The ranking of energy is the same in He^+ as it is in H.

(a) $n(5s) > n(4d)$
(b) $n(2s) = n(2p)$
(c) $n(6s) > n(3d)$
(d) $n(6s) = n(6p)$

S5-5. The $n = 7$ shell contains 49 spatial orbitals:

	n	ℓ	m_ℓ
7*s*	7	0	0
7*p*	7	1	−1 0 1
7*d*	7	2	−2 −1 0 1 2
7*f*	7	3	−3 −2 −1 0 1 2 3
7*g*	7	4	−4 −3 −2 −1 0 1 2 3 4
7*h*	7	5	−5 −4 −3 −2 −1 0 1 2 3 4 5
7*i*	7	6	−6 −5 −4 −3 −2 −1 0 1 2 3 4 5 6

In general, the nth shell supports n^2 combinations of n, ℓ, m_ℓ. Note that some authors skip the letter *i* in the enumeration of ℓ, proceeding from *h* ($\ell = 5$) to *j* ($\ell = 6$).

S5-6. The allowed values of m_ℓ range from $-\ell$ to $+\ell$ in steps of 1, creating $2\ell + 1$ possibilities:

$$m_\ell = 0, \pm 1, \pm 2, \ldots, \pm \ell$$

The h orbitals ($\ell = 5$) are therefore divided into 11 magnetic sublevels:

$$m_\ell = 0,\ \pm 1,\ \pm 2,\ \pm 3,\ \pm 4,\ \pm 5$$

S5-7. A $7g$ orbital has the quantum numbers $n = 7$ and $\ell = 4$:

Total number of nodes $= n - 1 = 6$

Total number of angular nodes $= \ell = 4$

Total number of radial nodes $= n - 1 - \ell = 2$

S5-8. The general designation is $n\ell$, where ℓ is assigned the code letters s (for $\ell = 0$), p (for $\ell = 1$), d (for $\ell = 2$), f (for $\ell = 3$), g (for $\ell = 4$), and so forth.

	SUBSHELL	n	ℓ
(a)	$5g$	5	4
(b)	$7s$	7	0
(c)	$6p$	6	1
(d)	$9d$	9	2

S5-9. Each subshell $n\ell$ supports $2\ell + 1$ values of m_ℓ, spanning the range $0, \pm 1, \pm 2, \ldots, \pm \ell$.

	SUBSHELL	n	ℓ	m_ℓ
(a)	$5g$	5	4	$0,\ \pm 1,\ \pm 2,\ \pm 3,\ \pm 4$
(b)	$7s$	7	0	0
(c)	$6p$	6	1	$0,\ \pm 1$
(d)	$9d$	9	2	$0,\ \pm 1,\ \pm 2$

S5-10. Each combination violates the requirement $0 \leq \ell < n$.

	SUBSHELL	n	ℓ	COMMENT
(a)	$2f$	2	3	forbidden ($\ell > n$)
(b)	$4g$	4	4	forbidden ($\ell = n$)
(c)	$3h$	3	5	forbidden ($\ell > n$)
(d)	$1p$	1	1	forbidden ($\ell = n$)

S5-11. Angular momentum quantum numbers $\ell = 0, 1, 2, 3, 4, 5, \ldots$ correspond to the letters *s*, *p*, *d*, *f*, *g*, *h*, ….

	n	ℓ	SUBSHELL
(a)	7	5	$7h$
(b)	4	2	$4d$
(c)	6	0	$6s$
(d)	5	1	$5p$

S5-12. The quantization rules are as follows:

$$n = 1, 2, 3, \ldots, \infty$$

$$\ell = 0, 1, 2, \ldots, n-1$$

$$m_\ell = 0, \pm 1, \pm 2, \ldots, \pm \ell$$

$$m_s = \pm \frac{1}{2}$$

Violations are cited in the table below:

	n	ℓ	m_ℓ	m_s	COMMENT
(a)	-1	0	0	$\frac{1}{2}$	forbidden ($n < 1$)
(b)	1	1	0	$\frac{1}{2}$	forbidden ($\ell = n$)
(c)	5	4	5	$-\frac{1}{2}$	forbidden ($\lvert m_\ell \rvert > \ell$)
(d)	5	5	4	$-\frac{1}{2}$	forbidden ($\ell = n$)

S5-13. Similar to the preceding exercise.

	n	ℓ	m_ℓ	m_s	COMMENT
(a)	1	0	0	$\frac{1}{4}$	forbidden ($m_s \neq \pm\frac{1}{2}$)
(b)	0	1	0	$\frac{1}{2}$	forbidden ($n < 1$)
(c)	3	2	-3	$-\frac{1}{2}$	forbidden ($\lvert m_\ell \rvert > \ell$)
(d)	$\frac{3}{2}$	$\frac{1}{2}$	$\frac{1}{2}$	$\frac{1}{2}$	forbidden ($n \neq 1, 2, \ldots, \infty$)

In (d), the quantum numbers $\ell = \frac{1}{2}$ and $m_\ell = \frac{1}{2}$ are also forbidden.

S5-14. See Exercise S5-12 for a summary of the quantization rules.

	n	ℓ	m_ℓ	m_s	COMMENT
(a)	1	0	0	1	forbidden ($m_s \neq \pm\frac{1}{2}$)
(b)	1	1	0	0	forbidden ($n = \ell,\ m_s \neq \pm\frac{1}{2}$)
(c)	3	2	−2	$-\frac{1}{2}$	allowed
(d)	10^4	1	0	$\frac{1}{2}$	allowed

S5-15. According to quantum mechanics, an electron has no definite path—no fixed orbit as stipulated by the Bohr theory. Instead, the electron is described probabilistically by a wave function $\psi(x, y, z)$. The square of the wave function, $\psi^2(x, y, z)$, is proportional to the probability of finding the electron in the vicinity of a given point (x, y, z).

S5-16. Inspection of the formula

$$r_n = \frac{\varepsilon_0 n^2 h^2}{\pi Z e^2 m_e} = \frac{n^2}{Z} a_0$$

enables us to identify the *Bohr radius*, a_0:

$$a_0 = \frac{\varepsilon_0 h^2}{\pi e^2 m_e}$$

(a) Insert the physical constants ε_0, h, e, and m_e into the expression for a_0:

$$a_0 = \frac{\left(8.854187817 \times 10^{-12}\ \text{C}^2\ \text{N}^{-1}\ \text{m}^{-2}\right)\left(6.6260755 \times 10^{-34}\ \text{J s}\right)^2}{\pi\left(1.60217733 \times 10^{-19}\ \text{C}\right)^2\left(9.1093897 \times 10^{-31}\ \text{kg}\right)} = 5.2917725 \times 10^{-11}\ \text{m}$$

The resulting value is approximately 0.529 Å.

Note that we use the following definitions to express the radius in meters:

$$1\ \text{N} = 1\ \text{kg m s}^{-2}$$

$$1\ \text{J} = 1\ \text{kg m}^2\ \text{s}^{-2}$$

(b) Substitute $n = 1$ and $Z = 1$ into the equation $r_n = (n^2/Z)a_0$ to obtain r_1 for the H atom:

$$r_1 = \frac{1^2}{1} a_0 = 5.29 \times 10^{-11}\ \text{m}$$

(c) Substitute $n = 2$ and $Z = 1$ to obtain $r_2 = 4a_0$ for hydrogen:

$$r_2 = \frac{2^2}{1}a_0 = 2.12 \times 10^{-10} \text{ m}$$

(d) Again. Substitute $n = 3$ and $Z = 1$ to obtain $r_3 = 9a_0$ for hydrogen:

$$r_3 = \frac{3^2}{1}a_0 = 4.76 \times 10^{-10} \text{ m}$$

S5-17. See the preceding exercise for the evaluation of r_n:

$$r_n = \frac{n^2}{Z}a_0$$

$$a_0 = \frac{\varepsilon_0 h^2}{\pi e^2 m_e} = 5.2917725 \times 10^{-11} \text{ m}$$

The increased nuclear charge in He^+ ($Z = 2$) causes the orbits to contract relative to those in the hydrogen atom. Each radius is half the value found in H:

Z	n	r_n
2	1	$\frac{1}{2}a_0 = 2.65 \times 10^{-11}$ m
2	2	$2a_0 = 1.06 \times 10^{-10}$ m
2	3	$\frac{9}{2}a_0 = 2.38 \times 10^{-10}$ m

S5-18. Similar, with $Z = 3$:

Z	n	r_n
3	1	$\frac{1}{3}a_0 = 1.76 \times 10^{-11}$ m
3	2	$\frac{4}{3}a_0 = 7.06 \times 10^{-11}$ m
3	3	$3a_0 = 1.59 \times 10^{-10}$ m

S5-19. Substitute the quantized velocity

$$v_n = \frac{Ze^2}{2\varepsilon_0 nh}$$

into the defining equation for kinetic energy:

$$E_{\mathrm{k},n} = \frac{1}{2} m_{\mathrm{e}} (v_n)^2 = \frac{1}{2} m_{\mathrm{e}} \left(\frac{Ze^2}{2\varepsilon_0 nh} \right)^2 = \frac{m_{\mathrm{e}} Z^2 e^4}{8\varepsilon_0^2 n^2 h^2}$$

S5-20. We carry over the quantized kinetic energy derived just above, adding to it the classical electrostatic energy between electron and nucleus:

$$E_{\mathrm{pot}} = -\frac{Ze^2}{4\pi\varepsilon_0 r}$$

Introduction of the quantized radius

$$r_n = \frac{\varepsilon_0 n^2 h^2}{\pi Z e^2 m_{\mathrm{e}}}$$

leads directly to an equation for the total energy of a one-electron atom.

(a) The total energy E_n is equal to the sum of the kinetic and potential contributions:

$$\begin{aligned} E_n &= E_{\mathrm{k},n} + E_{\mathrm{pot}} \\ &= \frac{m_{\mathrm{e}} Z^2 e^4}{8\varepsilon_0^2 n^2 h^2} - \frac{Ze^2}{4\pi\varepsilon_0 r_n} = \frac{m_{\mathrm{e}} Z^2 e^4}{8\varepsilon_0^2 n^2 h^2} - \frac{Ze^2}{4\pi\varepsilon_0} \times \frac{\pi Z e^2 m_{\mathrm{e}}}{\varepsilon_0 n^2 h^2} \\ &= \frac{m_{\mathrm{e}} Z^2 e^4}{8\varepsilon_0^2 n^2 h^2} - \frac{m_{\mathrm{e}} Z^2 e^4}{4\varepsilon_0^2 n^2 h^2} = -\frac{m_{\mathrm{e}} Z^2 e^4}{8\varepsilon_0^2 n^2 h^2} \end{aligned}$$

(b) Comparing the equations

$$E_n = -\frac{m_{\mathrm{e}} Z^2 e^4}{8\varepsilon_0^2 n^2 h^2} \qquad \text{and} \qquad E_n = -\frac{Z^2}{n^2} R_\infty$$

we identify the Rydberg constant, R_∞:

$$R_\infty = \frac{m_{\mathrm{e}} e^4}{8\varepsilon_0^2 h^2} = \frac{(9.1093897 \times 10^{-31}\ \mathrm{kg})(1.60217733 \times 10^{-19}\ \mathrm{C})^4}{8(8.854187817 \times 10^{-12}\ \mathrm{C^2\ N^{-1}\ m^{-2}})^2 (6.6260755 \times 10^{-34}\ \mathrm{J\ s})^2}$$

$$= 2.1798741 \times 10^{-18}\ \mathrm{J}$$

From the definition

$$1 \text{ J} = 1 \text{ N m} = 1 \text{ kg m}^2 \text{ s}^{-2}$$

we observe that R_∞ reduces properly to units of joules:

$$R_\infty \sim \frac{\text{kg C}^4}{\left(\text{C}^4 \text{ N}^{-2} \text{ m}^{-4}\right)\left(\text{J}^2 \text{ s}^2\right)} = \frac{\text{kg N}^2 \text{ m}^4}{\text{J}^2 \text{ s}^2} = \frac{\text{kg N}^2 \text{ m}^4}{\text{N}^2 \text{ m}^2 \text{ s}^2} = \text{kg m}^2 \text{ s}^{-2} = \text{J}$$

The Bohr theory of the one-electron atom yields precisely the same quantized energies as does the Schrödinger equation.

S5-21. If the mass of an electron were doubled (from m_e to $m_e' = 2m_e$), then both the Rydberg constant and the quantized energies would double as well:

$$R_\infty' = \frac{m_e' e^4}{8\varepsilon_0^2 h^2} = 2 \times \frac{m_e e^4}{8\varepsilon_0^2 h^2} = 2R_\infty$$

$$E_n' = -\frac{Z^2}{n^2} R_\infty' = -2 \times \frac{Z^2}{n^2} R_\infty = 2E_n$$

Since $E = hc/\lambda$, the wavelengths would be cut in half:

$$\lambda_{3\to2} = \frac{656 \text{ nm}}{2} = 328 \text{ nm}$$

$$\lambda_{4\to2} = \frac{486 \text{ nm}}{2} = 243 \text{ nm}$$

$$\lambda_{5\to2} = \frac{434 \text{ nm}}{2} = 217 \text{ nm}$$

$$\lambda_{6\to2} = \frac{410 \text{ nm}}{2} = 205 \text{ nm}$$

S5-22. Similar. If the charge of an electron were doubled (from e to $e' = 2e$), then the values

$$R'_\infty = \frac{m_e(e')^4}{8\varepsilon_0^2 h^2} = \frac{m_e(2e)^4}{8\varepsilon_0^2 h^2} = 16 \times \frac{m_e e^4}{8\varepsilon_0^2 h^2} = 16R_\infty$$

and

$$E'_n = -\frac{Z^2}{n^2} R'_\infty = -16 \times \frac{Z^2}{n^2} R_\infty = 16E_n$$

would each increase by a factor of 16. The Balmer wavelengths would decrease correspondingly:

$$\lambda_{3\to2} = \frac{656 \text{ nm}}{16} = 41.0 \text{ nm}$$

$$\lambda_{4\to2} = \frac{486 \text{ nm}}{16} = 30.4 \text{ nm}$$

$$\lambda_{5\to2} = \frac{434 \text{ nm}}{16} = 27.1 \text{ nm}$$

$$\lambda_{6\to2} = \frac{410 \text{ nm}}{16} = 25.6 \text{ nm}$$

S5-23. Ionization corresponds to a transition from $n_i = 1$ (the ground state) to $n_f = \infty$ (zero potential energy):

$$\Delta E = -R_\infty Z^2\left(\frac{1}{n_f^2} - \frac{1}{n_i^2}\right) = -R_\infty Z^2\left(\frac{1}{\infty^2} - \frac{1}{1^2}\right) = R_\infty Z^2 \equiv I$$

(a) See Exercise S5-21. If the mass of an electron were doubled, then both R_∞ and I would double as well:

	Z	$I' = (2R_\infty)Z^2$
H	1	$2R_\infty = 4.36 \times 10^{-18}$ J
He^+	2	$8R_\infty = 1.74 \times 10^{-17}$ J

(b) See Exercise S5-22. If the charge of an electron were doubled, then both R_∞ and I would increase by a factor of 16:

	Z	$I' = (16R_\infty)Z^2$
H	1	$16R_\infty = 3.49 \times 10^{-17}$ J
He^+	2	$64R_\infty = 1.40 \times 10^{-16}$ J

Chapter 6

Periodic Properties of the Elements

S6-1. Count the total number of *s* electrons in each of the following atoms (inclusive of core and valence):

(a) Ta **(b)** Ag **(c)** Cs **(d)** K **(e)** In

S6-2. Count the total number of *p* electrons in each of the following atoms (inclusive of core and valence):

(a) Ar **(b)** I **(c)** W **(d)** Ra **(e)** Li

S6-3. Count the total number of *d* electrons in each of the following atoms (inclusive of core and valence):

(a) P **(b)** Ti **(c)** Sn **(d)** La **(e)** Co

S6-4. Count the total number of *f* electrons in each of the following atoms (inclusive of core and valence):

(a) La **(b)** Lu **(c)** Pt **(d)** Rf **(e)** Xe

S6-5. Name an atom that contains a partially filled *f* subshell.

S6-6. Name an element from the *d* block.

S6-7. Name an element that contains a filled valence *p* subshell.

S6-8. Name two elements that have an s^1 valence configuration.

S6-9. **(a)** Write the electron configuration for a hydride ion, H^-. **(b)** Write the electron configuration for a helium atom, He. **(c)** Which species is more reactive, H^- or He? Explain why.

S6-10. Write electron configurations for the following atoms:

(a) Tc **(b)** I **(c)** Ag **(d)** Rb **(e)** Zn

S6-11. Write electron configurations for the following atoms:

(a) Xe **(b)** Cs **(c)** Ca **(d)** Tl **(e)** Bi

S6-12. Identify the atom from the valence configuration stated:

(a) $7s^1$ **(b)** $5s^2$ **(c)** $4s^2 3d^{10} 4p^3$ **(d)** $1s^2$ **(e)** $2s^2 2p^1$

S6-13. Identify the atom from the valence configuration stated:

(a) $4s^2 3d^3$ **(b)** $3s^2 3p^6$ **(c)** $6s^1$ **(d)** $6s^2$ **(e)** $6s^2 5d^1$

S6-14. Which of the following atoms are paramagnetic?

(a) He **(b)** Cs **(c)** Sr **(d)** Br **(e)** Ga

S6-15. Count the number of unpaired electrons in each atom:

(a) Ge **(b)** Kr **(c)** Fr **(d)** P **(e)** Ti

S6-16. Which species are diamagnetic?

(a) H^- **(b)** H **(c)** Cl^- **(d)** Cl **(e)** Ar

S6-17. Which species are diamagnetic?

(a) Mg^+ **(b)** Ca^{2+} **(c)** Li^+ **(d)** Al^+ **(e)** Sc

S6-18. Show what the periodic table would look like if the transition metals were separated from its main body—like the lanthanide and actinide elements.

S6-19. Show what the periodic table would look like if the lanthanide and actinide series were incorporated into its main body—like the transition metals.

S6-20. Write the electron configuration of element 119, as yet undiscovered. Where would it fall in the periodic table?

S6-21. To which period of the periodic table does the element europium belong—fourth, fifth, sixth, or seventh?

S6-22. To which period of the periodic table does the element americium belong—fourth, fifth, sixth, or seventh?

S6-23. How many elements would an eighth row of the periodic table contain?

S6-24. Arrange the following atoms in order of increasing radius:

(a) Sr, F, Li, Rb
(b) Al, Cl, H, Cs
(c) Ca, K, Fr, Si
(d) Se, F, Rn, Xe

S6-25. Arrange the following ions in order of increasing radius:

(a) Br^-, Li^+, Be^+, Be^{2+}
(b) Na^+, Mg^{2+}, Al^{3+}, Cl^-
(c) Br^-, I^-, O^{2-}, H^-
(d) Ca^{2+}, Sr^{2+}, Cs^+, Ba^{2+}

S6-26. Pick an atom, perhaps one of many, that fits each of the following descriptions:

(a) radius smaller than Br
(b) radius greater than Rn
(c) radius greater than K
(d) radius smaller than O

S6-27. Pick an atom, perhaps one of many, that fits each of the following descriptions:

(a) ionization energy greater than H
(b) ionization energy less than K
(c) ionization energy greater than P
(d) ionization energy less than Na

S6-28. Pick the species with the larger ionization energy:

(a) Li, He **(b)** Li, Li^+ **(c)** Li^+, He **(d)** Li^{2+}, He^+

S6-29. Arrange the following atoms in order of increasing ionization energy:

(a) Mg, Sr, He **(b)** F, Ba, Cl **(c)** Br, Rb, Li **(d)** Ne, Si, P

S6-30. Pick the atom with the more favorable electron affinity:

(a) Cs, Mg **(b)** Cs, Kr **(c)** Sb, I **(d)** I, Sr

S6-31. Explain why the alkaline earth metals have unfavorable electron affinities.

S6-32. Arrange the following elements (undiscovered as of the year 2000) in their expected order of **(a)** increasing ionization energy and **(b)** increasing atomic radius: 115, 117, 119.

S6-33. Why is helium assigned to Group VIII and not Group II?

S6-34. Noble gases are said to be inert, except in rare cases. If so, how do you account for the prevalence of neon signs? Isn't neon a noble gas?

SOLUTIONS

S6-1. An atom in period $n > 1$ accommodates $2(n - 1)$ s electrons in its $n - 1$ filled core shells. Group I elements (the alkali metals and hydrogen) add one valence s electron to this total, whereas all other representative elements (Groups II through VIII) add two. Most of the transition metals also have two valence s electrons, except in cases of anomalous configurations (for example, Ag ≡ [Kr]$5s^1 4d^{10}$).

					s ELECTRONS		
	ATOM	*Z*	PERIOD	GROUP	CORE	VALENCE	TOTAL
(a)	Ta	73	6	tm	10	2	12
(b)	Ag	47	5	tm	8	1	9
(c)	Cs	55	6	I	10	1	11
(d)	K	19	4	I	6	1	7
(e)	In	49	5	III	8	2	10

Key: tm = transition metal

S6-2. An atom in period $n > 2$ accommodates $6(n - 2)$ p electrons in core shells 2 through $n - 1$. Elements in Groups III through VIII add, in succession, one through six valence p electrons to this total. The alkali metals, the alkaline earth metals, and the transition metals have no p electrons in their valence shells.

	ATOM	Z	PERIOD	GROUP	p ELECTRONS CORE	VALENCE	TOTAL
(a)	Ar	18	3	VIII	6	6	12
(b)	I	53	5	VII	18	5	23
(c)	W	74	6	tm	24	0	24
(d)	Ra	88	7	II	30	0	30
(e)	Li	3	2	I	0	0	0

Key: tm = transition metal

S6-3. Atoms for which $Z \leq 20$ have no d electrons in their ground-state configurations. Starting with the fifth period ($n = 5$), the alkali metals (Group I), the alkaline earth metals (Group II), and the transition metals all accommodate $10(n - 4)$ d electrons in their core shells. Elements in Groups III through VIII acquire $10(n - 3)$ core d electrons, beginning in the fourth period.

The representative elements (Groups I through VIII) have no d electrons in valence orbitals. Transition metals in periods 4 through 7 contain between 1 and 10 valence d electrons. Lanthanide and actinide elements (sometimes called the *inner transition metals*) contain between 0 and 2 valence d electrons.

Consult Table C-8 in Appendix C of *Principles of Chemistry* for valence configurations of elements in the d block.

	ATOM	Z	PERIOD	GROUP	d ELECTRONS CORE	VALENCE	TOTAL
(a)	P	15	3	V	0	0	0
(b)	Ti	22	4	tm	0	2	2
(c)	Sn	50	5	IV	20	0	20
(d)	La	57	6	tm	20	1	21
(e)	Co	27	4	tm	0	7	7

Key: tm = transition metal

S6-4. Atoms for which $Z \leq 57$ have no f electrons in their ground states. Elements in the lanthanide and actinide series contain up to 14 valence f electrons. Consult Table C-8 in *Principles of Chemistry* for specific configurations.

Elements for which $72 \leq Z \leq 103$ contain 14 core f electrons. Beginning with rutherfordium ($Z = 104$), elements in period 7 contain 28 core electrons in $4f$ and $5f$ orbitals. See the table that follows.

	ATOM	Z	PERIOD	GROUP	f ELECTRONS CORE	VALENCE	TOTAL
(a)	La	57	6	tm	0	0	0
(b)	Lu	71	6	lan	0	14	14
(c)	Pt	78	6	tm	14	0	14
(d)	Rf	104	7	tm	28	0	28
(e)	Xe	54	5	VIII	0	0	0

Key: tm = transition metal lan = lanthanide series (inner transition metal)

S6-5. Cerium, first of the lanthanides, has the configuration $[Xe]6s^2 4f^1 5d^1$. Partially filled f orbitals are also found among the actinides, beginning with protactinium. In general, elements with $58 \leq Z \leq 69$ and $91 \leq Z \leq 101$ contain between 1 and 13 f electrons in an open shell.

S6-6. The d block begins with scandium ($Z = 21$), first of the fourth-row transition metals. Membership in the d block includes the elements scandium through zinc ($21 \leq Z \leq 30$), yttrium through cadmium ($39 \leq Z \leq 48$), lanthanum through mercury (excluding the lanthanides), and actinium through element 112 (excluding the actinides).

S6-7. Any one of the noble gases beginning with neon—Ne, Ar, Kr, Xe, Rn, and the newly synthesized element 118—contains a filled valence p subshell.

S6-8. Any of the elements in Group I—H, Li, Na, K, Rb, Cs, and Fr—has an s^1 valence configuration.

S6-9. The species H^- and He, although isoelectronic, exhibit drastically different chemical properties.

(a) $[H{:}]^-$ $1s^2$

(b) He: $1s^2$

(c) Neither system is readily able to accept an electron, since the first vacancy is in a $2s$ orbital of relatively high energy. The hydride ion, however, is more reactive than the noble gas helium. With a nuclear charge of only +1 to attract its two electrons, H^- can donate the electron pair and thereby function as a Lewis base.

S6-10. Orbitals fill in the usual order for $Z \leq 54$:

$$1s < 2s < 2p < 3s < 3p < 4s < 3d < 4p < 5s < 4d < 5p$$

(a) Tc $[Kr]5s^2 4d^5$

(b) I $[Kr]5s^2 4d^{10} 5p^5$

(c) Ag $[Kr]5s^1 4d^{10}$

(d) Rb $[Kr]5s^1$

(e) Zn $[Ar]4s^2 3d^{10}$

S6-11. The orbitals fill in the following sequence for $Z \leq 86$:

$$1s < 2s < 2p < 3s < 3p < 4s < 3d < 4p < 5s < 4d < 5p < 6s < 4f < 5d < 6p$$

(a) Xe $[Kr]5s^2 4d^{10} 5p^6$

(b) Cs $[Xe]6s^1$

(c) Ca $[Ar]4s^2$

(d) Tl $[Xe]6s^2 4f^{14} 5d^{10} 6p^1$

(e) Bi $[Xe]6s^2 4f^{14} 5d^{10} 6p^3$

S6-12. Consult the periodic table to establish the identity of each atom.

(a) The alkali metal *francium*, with a $7s^1$ valence configuration, appears as the first element of the seventh period: Fr, $Z = 87$.

(b) A $5s^2$ valence configuration tells us to locate the second element in period 5: the alkaline earth metal *strontium* (Sr, $Z = 38$).

(c) The atom is *arsenic* (As, $Z = 33$). First, the maximum value of n in the valence configuration $4s^2 3d^{10} 4p^3$ corresponds to a principal quantum number of 4 and hence period 4. Second, the three p electrons in $4p^3$ indicate that the atom lies three positions into the p block. Starting just after zinc ($4s^2 3d^{10}$), we count to 3: gallium, germanium, *arsenic*.

(d) *Helium*, the second element, is the only neutral species with the valence configuration $1s^2$.

(e) Locate *boron* (B, $Z = 5$) three positions into the second period, consistent with the valence configuration $2s^2 2p^1$. The full ground-state configuration is $1s^2 2s^2 2p^1$.

S6-13. The method is the same as in the preceding exercise.

	VALENCE CONFIGURATION	ATOM	Z	COMMENT
(a)	$4s^2 3d^3$	V	23	third element in period 4 of the *d* block
(b)	$3s^2 3p^6$	Ar	18	noble gas that completes period 3
(c)	$6s^1$	Cs	55	first element (Group I) of period 6
(d)	$6s^2$	Ba	56	second element (Group II) of period 6
(e)	$6s^2 5d^1$	La	57	first element in period 6 of the *d* block

S6-14. A paramagnetic atom contains one or more unpaired valence electrons. Observe that configurations ns^1 are paramagnetic, whereas configurations ns^2 are diamagnetic. Partially filled *p* subshells are populated according to Hund's rule.

	ATOM	Z	CONFIGURATION	SPIN ALIGNMENT	PARAMAGNETISM
(a)	He	2	$1s^2$	↑↓	no
(b)	Cs	55	$[Xe]6s^1$	↑	yes
(c)	Sr	38	$[Kr]5s^2$	↑↓	no
(d)	Br	35	$[Ar]4s^2 3d^{10} 4p^5$	↑↓ ↑↓ ↑↓ ↑ ($4s^2 4p^5$)	yes
(e)	Ga	31	$[Ar]4s^2 3d^{10} 4p^1$	↑↓ ↑ __ __ ($4s^2 4p^1$)	yes

For simplicity, in (d) and (e) we omit the filled d^{10} subshells from the spin diagrams.

S6-15. Similar to the preceding exercise: a count of unpaired electrons.

	ATOM	Z	CONFIGURATION	SPIN ALIGNMENT	NUMBER OF UNPAIRED e^-
(a)	Ge	32	$[Ar]4s^2 3d^{10} 4p^2$	↑↓ ↑ ↑ __ ($4s^2 4p^2$)	2
(b)	Kr	36	$[Ar]4s^2 3d^{10} 4p^6$	↑↓ ↑↓ ↑↓ ↑↓ ($4s^2 4p^6$)	0
(c)	Fr	87	$[Rn]7s^1$	↑	1
(d)	P	15	$[Ne]3s^2 3p^3$	↑↓ ↑ ↑ ↑	3
(e)	Ti	22	$[Ar]4s^2 3d^2$	↑↓ ↑ ↑ __ __ __	2

Core d^{10} configurations are omitted from the spin diagrams.

S6-16. A diamagnetic system, in contrast to a paramagnetic system, contains no unpaired electrons. The total number of electrons (N_e) is given by $Z - q$, where q is the signed charge on the species. For example, N_e is equal to $Z + q$ for an anion X^{q-}.

Configurations ns^2 and $ns^2 np^6$ are diamagnetic, whereas configurations ns^1 and $ns^2 np^j$ ($j \leq 5$) are paramagnetic. Partially filled *p* subshells are populated according to Hund's rule.

	SPECIES	N_e	CONFIGURATION	SPIN ALIGNMENT	DIAMAGNETISM
(a)	H^-	2	$1s^2$	↑↓	yes
(b)	H	1	$1s^1$	↑	no
(c)	Cl^-	18	$[Ne]3s^2 3p^6$	↑↓ ↑↓ ↑↓ ↑↓	yes
(d)	Cl	17	$[Ne]3s^2 3p^5$	↑↓ ↑↓ ↑↓ ↑	no
(e)	Ar	18	$[Ne]3s^2 3p^6$	↑↓ ↑↓ ↑↓ ↑↓	yes

H^- is isoelectronic with He. Cl^- is isoelectronic with Ar.

S6-17. Use the same method as in the preceding exercise. A cation X^{q+} contains a total of $Z - q$ electrons.

	SPECIES	N_e	CONFIGURATION	SPIN ALIGNMENT	DIAMAGNETISM
(a)	Mg^+	11	$[Ne]3s^1$	↑	no
(b)	Ca^{2+}	18	$[Ne]3s^2 3p^6$	↑↓ ↑↓ ↑↓ ↑↓	yes
(c)	Li^+	2	$1s^2$	↑↓	yes
(d)	Al^+	12	$[Ne]3s^2$	↑↓	yes
(e)	Sc	21	$[Ar]4s^2 3d^1$	↑↓ ↑ _ _ _ _	no

S6-18. Compare Figure 6.1(a) with Figure 6.1(b) on the next page.

S6-19. Compare Figure 6.1(a) with Figure 6.1(c). Note that in some renderings of the periodic table, lutetium and lawrencium are assigned to the scandium group in place of lanthanum and actinium. The group then reads Sc, Y, Lu, Lr from top to bottom, as is evident in Figure 6.1(c). See W. B. Jensen, *J. Chem. Educ.* **59**, 634–636 (1982) for the chemical arguments that support this choice.

S6-20. Element 118, with filled $6d$ and $7p$ subshells, completes the seventh period. Element 119, lying directly below Fr in Group I, will initiate a new period—the eighth:

Element 118	$[Rn]7s^2 5f^{14} 6d^{10} 7p^6$
Element 119	$[118]8s^1$

S6-21. Europium, a lanthanide element, has the configuration $[Xe]6s^2 4f^7$. Like the other lanthanides, it is a member of the sixth period; its $6p$ orbitals remain unoccupied.

The lanthanide elements are separated from the body of the table for convenience only, in order to compress the entire display into one page. These 14 atoms between lanthanum and hafnium properly occupy the middle of a *continuous* row of 32 elements making up period 6: Cs, Ba, La, Ce through Lu, Hf through Rn. See Exercise S6-19.

(a)

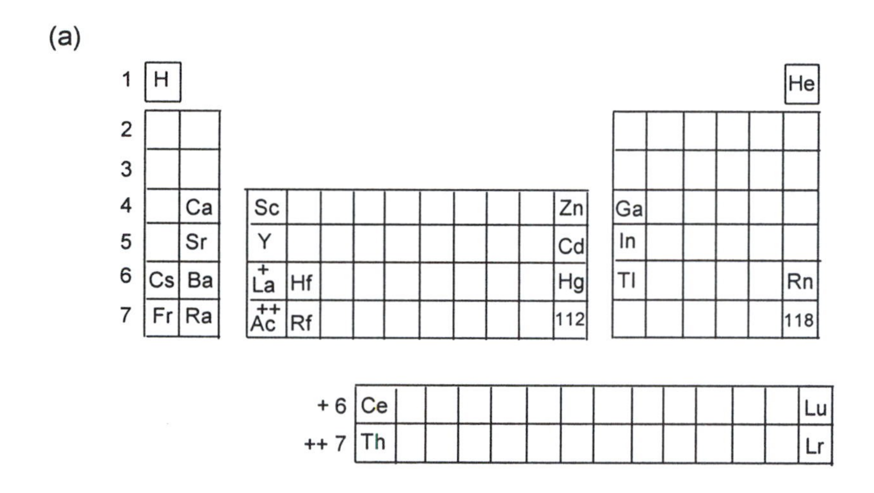

(b)

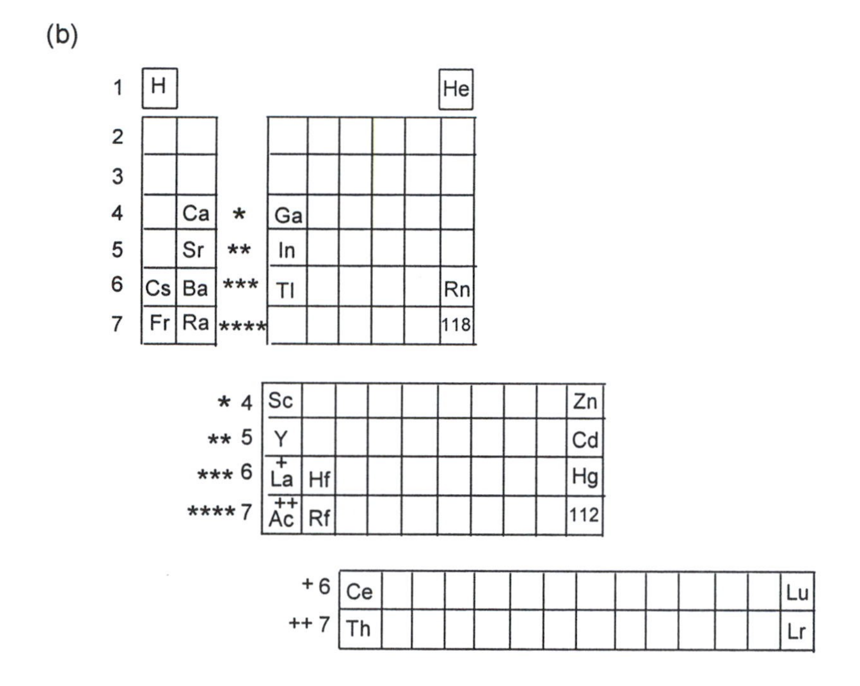

(c)

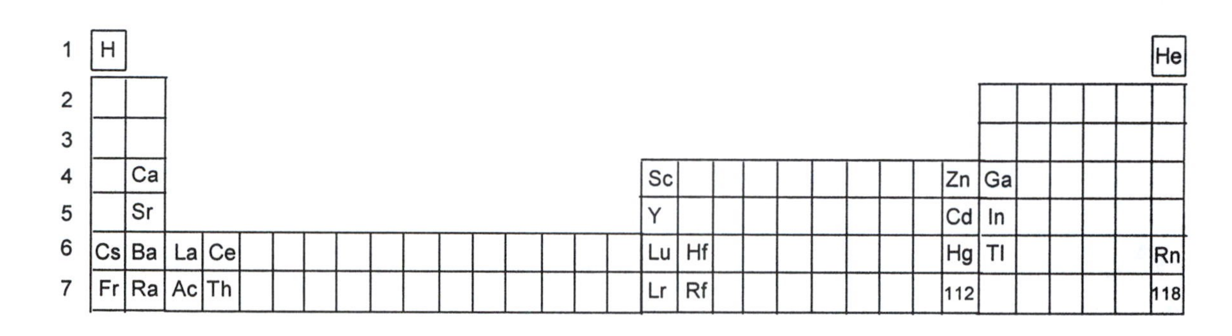

FIGURE 6.1 The periodic table. (a) Conventional representation. (b) An alternative form in which the *d* block is separated from the main-group elements. (c) A "long form" in which the *f*-block elements are incorporated into the body of the table.

S6-22. Similar. Americium, an actinide element, has the configuration $[Rn]7s^2 5f^7$. Like the other actinides, it belongs to the seventh period; its $7p$ orbitals remain unoccupied.

In an expanded version of the periodic table (Exercise S6-19), americium would be part of a continuous row of 32 atoms beginning with francium and ending with element 118.

S6-23. An eighth row of the periodic table would see the filling of the $8s$, $5g$, $6f$, $7d$, and $8p$ subshells:

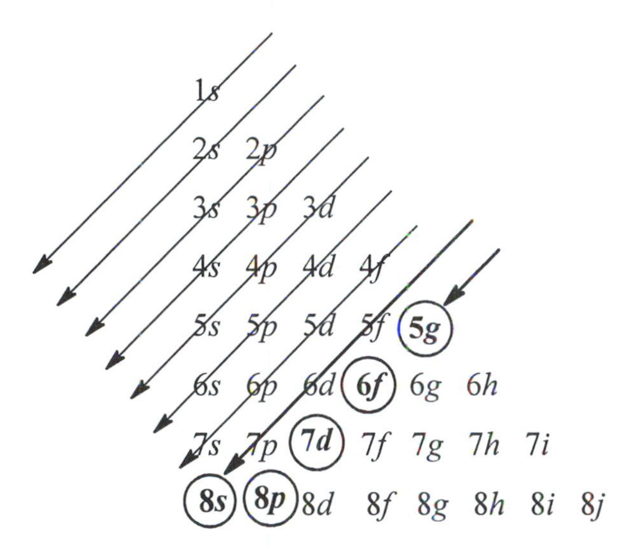

Running from element 119 (with configuration $[118]8s^1$) through element 168 (with configuration $[118]8s^2 5g^{18} 6f^{14} 7d^{10} 8p^6$, the new period would have 50 members:

s block: 2	*g* block: 18	*f* block: 14	*d* block: 10	*p* block: 6

S6-24. Atomic radius generally increases down a group and decreases across a period. Consult the periodic table to make the predictions stated below.

(a) Fluorine (Group VII) and lithium (Group I) lie at the far right and far left, respectively, of the second row. Strontium (Group II) and rubidium (Group I) both appear in the fifth row, far to the left:

$$R(\mathrm{F}) < R(\mathrm{Li}) < R(\mathrm{Sr}) < R(\mathrm{Rb})$$

(b) Hydrogen, the first element, is the smallest atom. Cesium, an alkali metal in period 6, is the largest of the four considered. Chlorine (Group VII) lies to the right of aluminum (Group III) in period 3:

$$R(\mathrm{H}) < R(\mathrm{Cl}) < R(\mathrm{Al}) < R(\mathrm{Cs})$$

(c) Silicon, in Group IV of period 3, is smallest. Next comes calcium (period 4, Group II), followed by potassium (period 4, Group I). Francium (period 7, Group I) is largest:

$$R(\mathrm{Si}) < R(\mathrm{Ca}) < R(\mathrm{K}) < R(\mathrm{Fr})$$

(d) Uppermost and far to the right (period 2, Group VII) is fluorine, smallest of the four atoms. Selenium, in Group VI of the fourth period, is both below and to the left of fluorine. Xenon terminates the fifth period, and radon terminates the sixth:

$$R(\text{F}) < R(\text{Se}) < R(\text{Xe}) < R(\text{Rn})$$

S6-25. Ions usually exhibit the same periodic trends as their parent atoms, with the following additional considerations: (1) A cation X^{q+}, bound by a comparatively high effective nuclear charge, has a smaller radius than its parent atom X. (2) A dipositive cation X^{2+} is bound more tightly than a monopositive cation X^{+} of the same atom. The ionic radius is smaller for the more highly charged species. (3) An anion X^{q-} has a larger radius than its neutral parent. (4) A doubly charged anion X^{2-} has a larger radius than a singly charged anion X^{-}. (5) If two ions are isoelectronic, then the one with the lower atomic number generally shows the larger radius.

(a) Beryllium, situated to the right of lithium in period 2, is the smaller atom and thus yields the smaller ion. Smallest of all three cations is doubly charged Be^{2+}, with four protons attracting only two electrons:

$$R(\text{Be}^{2+}) < R(\text{Be}^{+}) < R(\text{Li}^{+}) < R(\text{Br}^{-})$$

Largest of the set, in period 4, is the bromide anion, which has a radius greater than that of its parent atom.

(b) The cations Al^{3+}, Mg^{2+}, and Na^{+}, stripped to their cores and isoelectronic with neon, follow the same trend as the neutral parent atoms. Aluminum, with the highest atomic number and highest effective nuclear charge, has the smallest radius:

$$R(\text{Al}^{3+}) < R(\text{Mg}^{2+}) < R(\text{Na}^{+}) < R(\text{Cl}^{-})$$

The halide anion, again, is largest of the set.

(c) The hydride ion, considerably larger than a neutral hydrogen atom, proves also to be larger than the second-period oxide ion. Next comes the fourth-period bromide ion, followed by the fifth-period iodide ion:

$$R(\text{O}^{2-}) < R(\text{H}^{-}) < R(\text{Br}^{-}) < R(\text{I}^{-})$$

(d) The fourth-period calcium ion, lying directly above the strontium ion, has the smaller radius of the two. The barium and cesium ions, in period 6, are both larger species:

$$R(\text{Ca}^{2+}) < R(\text{Sr}^{2+}) < R(\text{Ba}^{2+}) < R(\text{Cs}^{+})$$

Cs^{+}, isoelectronic with Ba^{2+} but having the lower atomic number, is largest of all.

S6-26. Figure 6-13 on page 208 of *Principles of Chemistry* will be helpful for confirming the qualitative predictions made below. See also the data collected in Table C-8 of Appendix C, again in the text.

Remember, throughout, that atomic radii generally decrease across the rows and increase down the columns of the periodic table.

(a) Look for atoms situated higher in the periodic table than fourth-period bromine—but not too far toward the left. Suitable choices include P, S, Cl, and Ar in the third period, as well as B, C, N, O, F, and Ne in the second. Hydrogen and helium also qualify.

(b) Francium, which opens up the seventh period, clearly has a larger radius than radon, which closes the sixth. Other candidates include Na, Mg, K, Ca, Rb, Sr, Y, Cs, Ba, La, Ra, and many more—in general, elements toward the left and bottom of the table.

(c) Potassium, an alkali metal, has a relatively large atomic radius. Larger still, however, are the alkali metals that lie below it in the table: Rb, Cs, and Fr. All other atoms are smaller, although the alkaline earth metals Sr, Ba, and Ra are nearly the same size as K.

(d) Few atoms have a smaller radius than oxygen, which lies far to the right in the second period—well toward the top of the table. Only H and He, in the two-atom period just above, are significantly contracted relative to O. Fluorine, directly to the right in period 2, has just a slightly smaller radius.

S6-27. Ionization energies typically increase across the rows and decrease down the columns of the periodic table. See Figure 6-16 on page 212 of *Principles of Chemistry*, together with Table C-8 in Appendix C.

(a) Elements situated toward the right-hand side and top of the periodic table tend to have ionization energies greater than that of the hydrogen atom: helium, fluorine, and neon, for example. Other choices are possible as well.

(b) The circumstances are similar to those considered in Exercise S6-26(c). Alkali metals lying below potassium—Rb, Cs, Fr—have lower ionization energies.

(c) Atoms lying toward the right-hand side and top of the table tend to have ionization energies greater than that of third-period phosphorus. Selected examples include Cl and Ar in period 3, C through Ne in period 2, and both H and He in period 1.

(d) Similar to (b) above. The alkali metals lying below sodium (K, Rb, Cs, Fr) all have lower ionization energies.

S6-28. The applicable trends are as follows: (1) Of two atoms, the one located above and to the right generally has the larger ionization energy. (2) A cation is harder to ionize than its parent atom. (3) Of two isoelectronic species, the one with the higher atomic number usually has the greater ionization energy.

(a) $I(\text{He}) > I(\text{Li})$ trend 1

(b) $I(\text{Li}^+) > I(\text{Li})$ trend 2

(c) $I(\text{Li}^+) > I(\text{He})$ trend 3

(d) $I(\text{Li}^{2+}) > I(\text{He}^+)$ trend 3

S6-29. Again, consult the periodic table. Atoms situated toward the right-hand side (high effective nuclear charges) and top (low principal quantum numbers) tend to have high ionization energies.

(a) Strontium lies two rows below magnesium in Group II. Helium, a first-period noble gas, has the highest ionization energy of the three:

$$I(\text{Sr}) < I(\text{Mg}) < I(\text{He})$$

(b) Barium, an alkaline earth metal in the sixth period, is easiest to ionize. Second-period fluorine, positioned one row above chlorine, is hardest:

$$I(\text{Ba}) < I(\text{Cl}) < I(\text{F})$$

(c) The fourth-period halogen (Br), despite lying well below the second-period alkali metal (Li), has the largest ionization energy of the set:

$$I(\text{Rb}) < I(\text{Li}) < I(\text{Br})$$

The high effective nuclear charge in bromine offsets the relatively high principal quantum number.

(d) The ionization energies increase in the order Si (period 3, Group IV), P (period 3, Group V), Ne (period 2, Group VIII):

$$I(\text{Si}) < I(\text{P}) < I(\text{Ne})$$

S6-30. Periodic trends in electron affinity are analogous to those observed for radius and ionization energy, although not quite as regular. See Figure 6-17 on page 214 of *Principles of Chemistry*.

In the explanations below, the sign ">" should be interpreted as "more favorable than" (having a more negative electron affinity).

(a) Magnesium, with an s^2 valence configuration, has no affinity for an extra electron. By contrast, cesium is able to add a second electron to its s^1 configuration and thereby fill the open subshell:

$$EA(\text{Cs}) > EA(\text{Mg})$$

(b) Similar. Krypton is a noble gas. It has no low-lying vacant orbitals in which to accommodate an additional electron:

$$EA(\text{Cs}) > EA(\text{Kr})$$

(c) Iodine, only one electron shy of a p^6 configuration, is better able than antimony to add an electron:

$$EA(\text{I}) > EA(\text{Sb})$$

(d) See parts (a) and (c) above. With its s subshell already filled, the alkaline earth metal has little or no electron affinity. The halogen, however, realizes a closed-shell p^6 configuration upon attachment of an extra electron:

$$EA(\text{I}) > EA(\text{Sr})$$

S6-31. The ns^2 valence configuration of an alkaline earth metal (Group II) resists the addition of an extra electron. The effective nuclear charge (Z_{eff}) is low, and the atom would pay an energetic penalty to open up a new subshell.

S6-32. See Figure 6.1 on page 74 of this manual. Elements 115 and 117 will fall in the seventh period, with effective charge increasing from left to right across the row. Element 119, which will initiate a new period (the eighth) in the table, is expected to be an alkali metal with an $8s^1$ valence configuration.

(a) Element 119, with a principal quantum number of 8, should have a smaller ionization energy than either of the two elements for which $n = 7$. Within the seventh period itself, we expect I to be larger for the atom with the higher effective nuclear charge:

$$I(119) < I(115) < I(117)$$

(b) Atomic radii increase in the same order that ionization energies *decrease*:

$$R(117) < R(115) < R(119)$$

The atom with the highest principal quantum number, element 119, should have the largest radius.

S6-33. Despite having two valence electrons and an ns^2 configuration, helium shares little in common with the alkaline earth metals of Group II. Helium, energetically satisfied with its completed $n = 1$ shell, is unable to lose or gain electrons in ordinary chemical combination. It behaves as a noble gas and is logically assigned to Group VIII.

By contrast, the alkaline earth metals readily lose their two electrons to form dipositive cations isoelectronic with the preceding noble gas. The outer electrons are layered over a core of filled orbitals, and the effective nuclear charge is comparatively low.

S6-34. The glow of a neon sign arises from an electrical discharge. Stripped away by violent application of an external voltage, ionized electrons emit energy as they are recaptured by the cationic core species.

Such action does not result in the sharing of electrons or the formation of new compounds. Each atom, remaining chemically aloof, is affected individually and does not combine with other atoms.

Chapter 7

Covalent Bonding and Molecular Orbitals

S7-1. Suppose that a homonuclear diatomic system (second row) has an odd number of electrons. **(a)** Do you have sufficient information to determine whether the structure is a molecule or an ion? **(b)** If you were to determine that the structure is ionic, would you be able to tell whether it is an anion or a cation?

S7-2. Suppose that a heteronuclear diatomic system (second row) has an odd number of electrons. **(a)** Do you have sufficient information to determine whether the structure is a molecule or an ion? **(b)** If you were to determine that the structure is ionic, would you be able to tell whether it is an anion or a cation?

S7-3. Identify the homonuclear diatomic molecule (second row) that best fits each of the following descriptions:

(a) 6 bonding and 4 antibonding electrons (inclusive of core and valence)
(b) 4 bonding and 2 antibonding electrons (inclusive of core and valence)
(c) 2 bonding electrons (valence)
(d) 4 bonding and 2 antibonding electrons (valence)

S7-4. A homonuclear diatomic species (second row) has a bond order of 2.5. **(a)** Is the species a molecule or an ion? **(b)** Write a possible valence electron configuration. **(c)** Is your proposed species paramagnetic? If so, how many electrons are unpaired?

S7-5. Identify the homonuclear diatomic molecule (second row) that best fits each of the following descriptions:

(a) 4 bonding electrons (valence); bond order = 1; paramagnetic
(b) 6 antibonding electrons (valence); bond order = 1; diamagnetic
(c) 2 bonding electrons (valence); bond order = 1; diamagnetic

S7-6. Identify the homonuclear diatomic molecule (second row) that best fits each of the following descriptions:

(a) bond order = 2; paramagnetic
(b) bond order = 2; diamagnetic
(c) bond order = 3; diamagnetic

One of the two pieces of information given in part (c) is redundant. Which is it?

S7-7. Consider the following diatomic systems: Li_2^+, Li_2, Li_2^-.

(a) Calculate the bond order for each structure.
(b) Which structure has the largest dissociation energy?
(c) Which structure has the shortest bond length?

S7-8. Consider the following diatomic systems: Be_2^+, Be_2, Be_2^-.

(a) Calculate the bond order for each structure.
(b) Which structure has the smallest dissociation energy?
(c) Which structure has the longest bond length?

S7-9. Is it theoretically possible for a second-row, homonuclear diatomic system in its ground state to have a bond order equal to 4? If so, use an energy-level diagram to sketch a plausible electron configuration.

S7-10. Is it theoretically possible for a second-row, homonuclear diatomic system in an *excited* state to have a bond order equal to 4? If so, use an energy-level diagram to sketch a plausible electron configuration.

S7-11. **(a)** Explain the significance of bonding, antibonding, and nonbonding orbitals in the molecular orbital picture of chemical bonding. **(b)** Is the energy of a bonding molecular orbital greater than, less than, or equal to the energy of the two atomic orbitals from which it arises? **(c)** Same question, but for an antibonding orbital: Is the energy greater than, less than, or equal to the energy of the original atomic orbitals? **(d)** Once more, for a nonbonding molecular orbital: Is the energy greater than, less than, or equal to the energy of the atomic orbitals?

S7-12. Count the number of bonding, nonbonding, and antibonding electrons in the π system of each species:

(a) $CH_2CHCH_2^+$, the allyl cation
(b) CH_2CHCH_2, the allyl free radical
(c) $CH_2CHCH_2^-$, the allyl anion

S7-13. **(a)** Draw a Lewis structure for the molecule C_2Cl_2. **(b)** Identify each bond as σ or π. **(c)** Describe the bond angles and hybridization around each carbon. **(d)** Does the molecule have an overall dipole moment?

S7-14. Rank the following molecules in order of increasing dipole moment: C_2H_2, C_2Cl_2, C_2HF, C_2HCl.

S7-15. **(a)** Draw a Lewis structure for the molecule CH_3CH_2Br. **(b)** Identify each bond as σ or π. **(c)** Describe the bond angles and hybridization around each carbon.

S7-16. **(a)** Draw a Lewis structure for the molecule CH_2CHCH_3. **(b)** Identify each bond as σ or π. **(c)** Describe the bond angles and hybridization around each carbon.

S7-17. **(a)** Draw a Lewis structure for the molecule $CH_2CHCHCHCH_3$. **(b)** Identify each bond as σ or π. **(c)** Describe the bond angles and hybridization around each carbon.

S7-18. **(a)** Sketch the lowest delocalized π orbital for the molecule $CH_2CHCHCHCHCH_2$. **(b)** How many nodes does it contain? **(c)** Sketch the highest delocalized π orbital. **(d)** How many nodes does it contain?

S7-19. **(a)** Draw the Lewis structure of $BeCl_2$. **(b)** Use the VSEPR model to predict the shape of the molecule. **(c)** Is the dipole moment of the molecule greater than, less than, or the same as the dipole moment of BeF_2?

S7-20. **(a)** Use the VSEPR model to predict the geometry of the electron pairs in IF_5. How many bonding pairs and how many lone pairs are disposed about the central atom? **(b)** What is the likely shape of the molecule?

S7-21. **(a)** Draw three resonance structures for the nitrate ion, NO_3^-. **(b)** Use the VSEPR model to predict the shape of the ion. **(c)** Propose a hybridization scheme consistent with the geometry.

S7-22. The formate ion, HCO_2^-, is a planar structure with bond angles of approximately 120°. **(a)** Draw two possible resonance forms. **(b)** Propose a hybridization scheme consistent with the geometry. **(c)** Describe the π bond. Is it localized or delocalized?

S7-23. Consider again the formate ion, HCO_2^-. **(a)** Sketch the lowest π molecular orbital. **(b)** Predict whether the bond order of each C—O linkage will fall between 0 and 1, between 1 and 2, or between 2 and 3. Take into account both σ and π electrons.

S7-24. **(a)** Use an energy-level diagram to sketch a possible π-electron configuration for the benzene anion, $C_6H_6^-$. **(b)** Is the ion more stable or less stable relative to neutral C_6H_6?

S7-25. Suppose that a molecule contains 10 atoms and 50 electrons. Is it theoretically possible for the electron density between two particular atoms to have the value 0.0647? What does it mean to speak of "647 ten-thousandths of an electron"?

S7-26. Fluorine is the most electronegative of all the elements. Are all heteronuclear bonds involving fluorine highly polar? Explain why or why not.

SOLUTIONS

S7-1. Each atom A contributes n valence electrons to a homonuclear diatomic molecule A_2, creating a total of $2n$.

(a) Yes—a *homonuclear* diatomic system with an odd number of electrons can only be an ion, not a neutral molecule. The integer $2n$ is necessarily even for all values of n.

(b) No—without additional information, we cannot determine whether an odd-electron homonuclear structure carries a positive or negative charge. Diatomic anions and cations both may have an odd number of electrons.

S7-2. Atom A contributes n valence electrons, and atom B contributes m to a heteronuclear diatomic molecule AB. The total is $n + m$.

(a) No—a heteronuclear diatomic system with an odd number of valence electrons may be either neutral or ionic. If n and m are both even (or both odd), then the sum $n + m$ will be even. If one number is even and the other is odd, then the total will be odd. Neither outcome is ruled out for an ion or a neutral molecule.

(b) No—electrons may be added or subtracted to create odd-number anions, even-number anions, odd-number cations, and even-number cations.

S7-3. Valence molecular orbitals in homonuclear diatomic systems (second row) are filled according to the following sequences:

$$\sigma_{2s} < \sigma_{2s}^* < \pi_{2p} < \sigma_{2p_z} < \pi_{2p}^* < \sigma_{2p_z}^* \qquad (Z \leq 7)$$

$$\sigma_{2s} < \sigma_{2s}^* < \sigma_{2p_z} < \pi_{2p} < \pi_{2p}^* < \sigma_{2p_z}^* \qquad (Z \geq 8)$$

Hund's rule and the Pauli exclusion principle apply.

Below the valence levels are two core levels, both of which are fully occupied in a ground-state molecule:

$$\sigma_{1s} < \sigma_{1s}^*$$

(a) The description calls for six bonding and four antibonding electrons to be distributed over the core and valence orbitals. Since there are 10 electrons in all, we know that the atomic number of the atom A in A_2 must be 5. The molecule is B_2 (diatomic boron):

$$B_2 \qquad (\sigma_{1s})^2(\sigma_{1s}^*)^2(\sigma_{2s})^2(\sigma_{2s}^*)^2(\pi_{2p})^2$$

(b) Here we have four bonding and two antibonding electrons, inclusive of core and valence. Similar reasoning leads us to identify the structure as Li_2 ($Z = 3$):

$$Li_2 \qquad (\sigma_{1s})^2(\sigma_{1s}^*)^2(\sigma_{2s})^2$$

(c) Same as (b), except now we specify only the two bonding electrons in the σ_{2s} valence orbital:

$$Li_2 \qquad (\sigma_{2s})^2$$

(d) Same as (a), but with specification of only the six valence electrons (four bonding, two antibonding):

$$B_2 \qquad (\sigma_{2s})^2(\sigma_{2s}^*)^2(\pi_{2p})^2$$

S7-4. We are told only that some second-row diatomic structure has a bond order of 2.5.

(a) The species is an ion. The half-integral bond order implies an odd number of electrons—a telltale sign of an ionic system A_2^{q+} or A_2^{q-}, as explained in Exercise S7-1. Five more electrons, net, are in bonding orbitals than in antibonding orbitals.

(b) Acceptable choices, with selected examples, are shown below:

VALENCE CONFIGURATION	BOND ORDER = $\frac{5}{2}$	EXAMPLE
$(\sigma_{2s})^2(\sigma_{2s}^*)^2(\sigma_{2p_z})^2(\pi_{2p})^4(\pi_{2p}^*)^1$	$\frac{1}{2}(8-3)$	O_2^+
$(\sigma_{2s})^2(\sigma_{2s}^*)^2(\pi_{2p})^4(\sigma_{2p_z})^1$	$\frac{1}{2}(7-2)$	C_2^- N_2^+
$(\sigma_{2s})^2(\sigma_{2s}^*)^2(\pi_{2p})^4(\sigma_{2p_z})^2(\pi_{2p}^*)^1$	$\frac{1}{2}(8-3)$	N_2^-

(c) The odd electron makes each of the possible species paramagnetic. One electron remains unpaired, whether in a σ bonding orbital or in a π^* antibonding orbital.

S7-5. Bond order is defined as follows:

$$\text{Bond order} = \tfrac{1}{2}(\text{no. bonding e}^- - \text{no. antibonding e}^-)$$

All electrons are paired in a diamagnetic structure. A paramagnetic structure has one or more unpaired electrons.

(a) Given four bonding electrons (valence) and a bond order equal to 1, we determine that there are two antibonding electrons in valence orbitals:

$$\text{No. antibonding e}^- = (\text{no. bonding e}^-) - 2 \times (\text{bond order}) = 4 - 2(1) = 2$$

With a total of 10 electrons (six in the valence orbitals and, implicitly, four in the core), we then identify the homonuclear diatomic molecule as B_2 ($Z = 5$):

$$B_2 \qquad (\sigma_{1s})^2(\sigma^*_{1s})^2(\sigma_{2s})^2(\sigma^*_{2s})^2(\pi_{2p})^2 \qquad \text{(core + valence)}$$

$$B_2 \qquad (\sigma_{2s})^2(\sigma^*_{2s})^2(\pi_{2p})^2 \qquad \text{(valence only)}$$

The two π electrons occupy separate orbitals, their spins parallel in accordance with Hund's rule. B_2 is paramagnetic:

$\sigma^*_{2p_z}$	—
π^*_{2p}	— —
σ_{2p_z}	—
π_{2p}	↑ ↑
σ^*_{2s}	↑↓
σ_{2s}	↑↓

(b) The molecule is described as diamagnetic, with six antibonding valence electrons and a bond order equal to 1. If so, then there must be an additional eight bonding electrons:

$$\text{No. bonding e}^- = (\text{no. antibonding e}^-) + 2 \times (\text{bond order}) = 6 + 2(1) = 8$$

The valence configuration contains 14 electrons,

$$(\sigma_{2s})^2(\sigma_{2s}^*)^2(\sigma_{2p_z})^2(\pi_{2p})^4(\pi_{2p}^*)^4$$

and the total number of electrons, inclusive of core and valence, is 18. The molecule is fluorine (F_2, $Z = 9$). All the electrons are paired.

(c) With just two bonding electrons in valence orbitals, a bond order of 1 can be realized only if there are no offsetting antibonding electrons:

$$\text{No. antibonding } e^- = (\text{no. bonding } e^-) - 2 \times (\text{bond order}) = 2 - 2(1) = 0$$

The molecule is Li_2. Both electrons are paired in the σ_{2s} orbital.

S7-6. Configurations that contain one, two, or three electrons in π or π^* orbitals are paramagnetic:

↑ __ ↑ ↑ ↑↓ ↑

A four-electron π configuration is diamagnetic. All the electrons are paired:

↑↓ ↑↓

(a) The oxygen molecule, with two unpaired π^* electrons, is paramagnetic and has a bond order equal to 2:

$$O_2 \qquad (\sigma_{2s})^2(\sigma_{2s}^*)^2(\sigma_{2p_z})^2(\pi_{2p})^4(\pi_{2p}^*)^2 \quad \text{(paramagnetic)}$$

$$\text{Bond order} = \tfrac{1}{2}(8 - 4) = 2$$

We rule out diamagnetic C_2, which also has a bond order of 2. See part (b) immediately below.

(b) The π electrons in C_2 are all paired:

$$C_2 \qquad (\sigma_{2s})^2(\sigma_{2s}^*)^2(\pi_{2p})^4 \quad \text{(diamagnetic)}$$

$$\text{Bond order} = \tfrac{1}{2}(6 - 2) = 2$$

(c) N_2 is diamagnetic and has a bond order of 3:

$$N_2 \qquad (\sigma_{2s})^2(\sigma_{2s}^*)^2(\pi_{2p})^4(\sigma_{2p_z})^2 \quad \text{(diamagnetic)}$$

$$\text{Bond order} = \tfrac{1}{2}(8 - 2) = 3$$

We would have identified nitrogen as the mystery molecule even without being told that the structure is diamagnetic. No other homonuclear diatomic molecule has a bond order equal to 3.

S7-7. Count one, two, and three valence electrons in the cation, neutral molecule, and anion, respectively:

$$Li_2^+ \qquad (\sigma_{2s})^1$$

$$Li_2 \qquad (\sigma_{2s})^2$$

$$Li_2^- \qquad (\sigma_{2s})^2(\sigma_{2s}^*)^1$$

(a) Bond order is lower in the ions and higher in the molecule:

$$Li_2^+ \qquad \text{Bond order} = \tfrac{1}{2}(1 - 0) = \tfrac{1}{2}$$

$$Li_2 \qquad \text{Bond order} = \tfrac{1}{2}(2 - 0) = 1$$

$$Li_2^- \qquad \text{Bond order} = \tfrac{1}{2}(2 - 1) = \tfrac{1}{2}$$

(b) Li_2, the neutral molecule, has the highest bond order and consequently the largest dissociation energy of the three species considered.

(c) Strong bonds are short bonds. Li_2 is expected to have the shortest bond length.

S7-8. The method is the same as in Exercise S7-7.

(a) Fill the valence orbitals of the cation, neutral molecule, and anion with three, four, and five electrons, respectively:

$$Be_2^+ \qquad (\sigma_{2s})^2(\sigma_{2s}^*)^1 \qquad \text{Bond order} = \tfrac{1}{2}(2 - 1) = \tfrac{1}{2}$$

$$Be_2 \qquad (\sigma_{2s})^2(\sigma_{2s}^*)^2 \qquad \text{Bond order} = \tfrac{1}{2}(2 - 2) = 0$$

$$Be_2^- \qquad (\sigma_{2s})^2(\sigma_{2s}^*)^2(\pi_{2p})^1 \qquad \text{Bond order} = \tfrac{1}{2}(3 - 2) = \tfrac{1}{2}$$

(b) The neutral molecule, Be_2, has a bond order of 0 and is therefore unstable. Its dissociation energy, theoretically equal to zero, has been shown experimentally to be only 10 kJ mol^{-1}.

(c) For the same reason, Be_2 is expected to have the longest bond length of the three species (or, according to the simple model, not to form a bond at all).

S7-9. No. If orbitals are filled according to the Aufbau principle, then a second-row diatomic molecule cannot acquire a net excess of eight bonding electrons. Electrons occupying antibonding orbitals limit the maximum bond order to 3, realized only in N_2.

S7-10. Yes. A bond order of 4 is possible if the structure departs from the ground-state configuration prescribed by the Aufbau principle. For example, we can imagine a doubly excited C_2 molecule in which two σ^*_{2s} electrons are promoted to the σ_{2p_z} orbital:

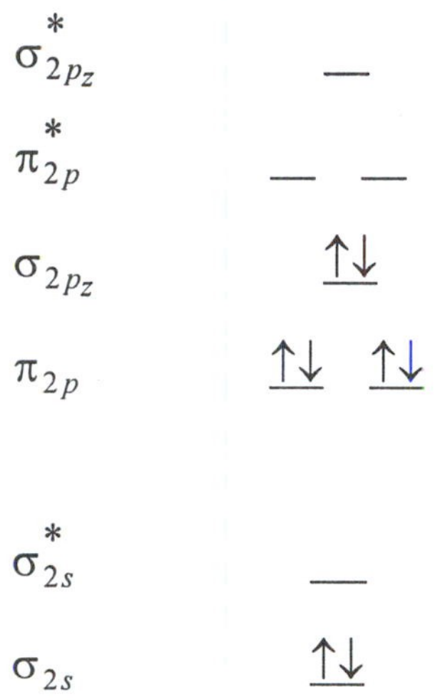

$\sigma^*_{2p_z}$	—
π^*_{2p}	— —
σ_{2p_z}	↑↓
π_{2p}	↑↓ ↑↓
σ^*_{2s}	—
σ_{2s}	↑↓

The eight bonding electrons are not offset by any antibonding electrons, and the resulting bond order is equal to 4:

$$\text{Bond order} = \tfrac{1}{2}(\text{no. bonding e}^- - \text{no. antibonding e}^-) = \tfrac{1}{2}(8 - 0) = 4$$

By contrast, the bond order of the ground-state configuration is only 2:

$$C_2 \qquad (\sigma_{2s})^2(\sigma^*_{2s})^2(\pi_{2p})^4$$

$$\text{Bond order} = \tfrac{1}{2}(6 - 2) = 2$$

S7-11. See Section 7-1 of the text for a conceptual overview of molecular orbitals.

(a) Electrons in a bonding orbital are concentrated between the nuclei to a greater extent than they would be in a noninteracting system. Electrons in an antibonding orbital are concentrated to a lesser extent. Electrons in a nonbonding orbital are concentrated neither more nor less than they would be in a noninteracting system.

(b) The energy of a bonding molecular orbital is lower than the energy of the separated atomic orbitals. Caught between the nuclei, a bonding electron is simultaneously attracted to more than one positive site.

(c) The energy of an antibonding orbital is higher than the energy of the separated atomic orbitals. The electron, to some extent, is deprived of the stabilizing influence of the nuclei.

(d) Electrons in a nonbonding orbital interact with only one nucleus. The energy is the same as it is for the separated atoms.

S7-12. Each of the three carbons contributes one electron to the π system of the neutral species. Loss of a single electron produces the cation; gain of an electron produces the anion:

	$CH_2CHCH_2^+$	CH_2CHCH_2	$CH_2CHCH_2^-$
π^*	—	—	—
π^n	—	↑	↑↓
π	↑↓	↑↓	↑↓

The π-electron count is summarized below:

SPECIES	TOTAL	BONDING	NONBONDING	ANTIBONDING
(a) $CH_2CHCH_2^+$	2	2	0	0
(b) CH_2CHCH_2	3	2	1	0
(c) $CH_2CHCH_2^-$	4	2	2	0

See Example 7-8 in *Principles of Chemistry.*

S7-13. Our molecule C_2Cl_2 is an analogue of triple-bonded acetylene, HC≡CH. The two carbon atoms contribute four valence electrons apiece, and the two chlorine atoms each contribute seven—22 valence electrons in all.

(a) Each carbon atom forms four bonds:

:C̈l—C≡C—C̈l:

(b) The C–Cl single bonds are σ linkages. The C≡C triple bond consists of one σ linkage and two π linkages. Three lone pairs remain on the terminal chlorine sites.

(c) VSEPR theory predicts a linear structure for the molecule, since effectively two sets of electron pairs appear on each of the two carbon sites. The resulting bond angles of 180° are consistent with *sp* hybridization.

(d) Although each C–Cl bond is individually polar, the bond dipoles are diametrically opposed in the linear molecule. They cancel, leaving the structure with no net dipole moment.

S7-14. Fluorine, more electronegative than chlorine, produces a larger dipole moment in the monosubstituted molecule:

H–C≡C–H = Cl–C≡C–Cl < H–C≡C–Cl < H–C≡C–F

Both H–C≡C–H and Cl–C≡C–Cl are nonpolar, as explained in the preceding exercise.

S7-15. The molecule CH_3CH_2Br is a brominated analogue of ethane, $H_3C–CH_3$.

(a) Twenty valence electrons contribute to the structure (four from each of two carbons, one from each of five hydrogens, and seven from bromine):

```
     H   H
     |   |      ..
 H—C—C—Br:
     |   |      ..
     H   H
```

Three lone pairs are localized on the bromine atom.

(b) Each of the seven connections, a single bond, is of σ symmetry.

(c) With four groups attached to each carbon, VSEPR theory predicts a tetrahedral geometry around these two sites. The bond angles should all be close to 109.5°, consistent with sp^3 hybridization.

S7-16. More practice with simple hydrocarbons.

(a) The molecule CH_2CHCH_3 contains 18 valence electrons. If each carbon atom is to form four bonds, then one of the connections must be a double bond:

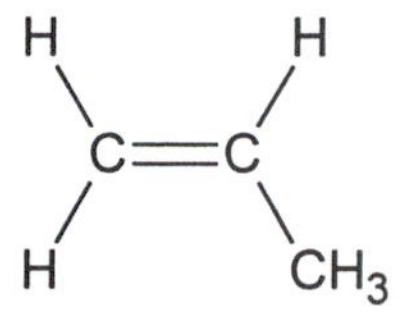

H H
C=C
H CH_3

(b) The C=C double bond consists of a σ linkage and a π linkage. The C–CH_3 single bond and all six C–H single bonds are of σ symmetry.

(c) Three sets of electrons surround each of the two double-bonded carbons. The VSEPR model prescribes a trigonal planar geometry (sp^2), with approximately 120° bond angles all around:

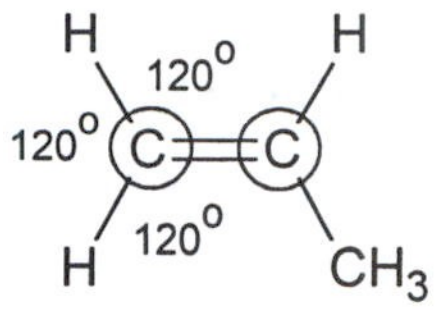

H 120° H
120° C=C
H 120° CH_3

The remaining carbon, hybridized as sp^3 in –CH_3, is tetrahedral. Bond angles are approximately 109.5°:

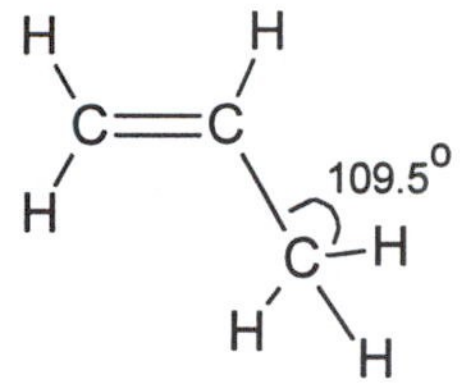

H H
C=C
H 109.5°
C–H
H H

S7-17. Here we have a molecule with a conjugated π system.

(a) There are 28 valence electrons in C_5H_8. Two double bonds enable carbon to have a valence of 4 at all positions:

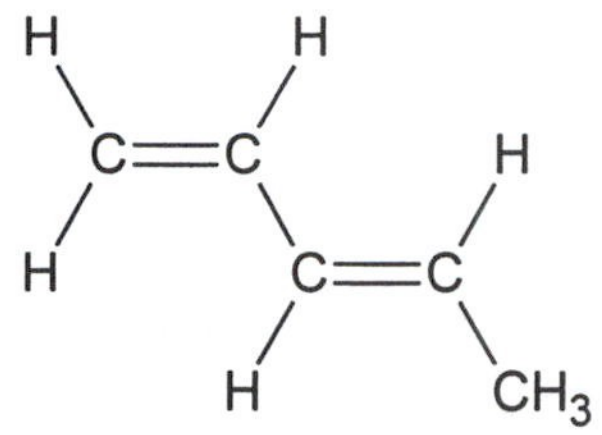

H H
C=C H
H C=C
H CH_3

(b) Each of the two C=C double bonds consists of a σ linkage and a π linkage. The remaining C–C and C–H single bonds are σ linkages.

(c) The C=C carbons are hybridized as sp^2, appropriate for trigonal planar geometry (≈120° bond angles). The CH_3 carbon is hybridized tetrahedrally as sp^3 (≈109.5° bond angles), similar to the structure in the preceding exercise.

S7-18. Each carbon contributes one unhybridized *p* orbital to an unbroken π system:

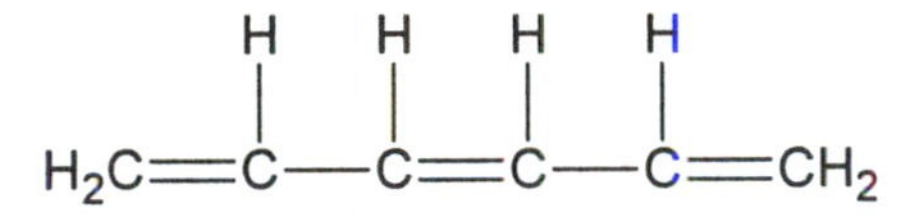

For simplicity, we represent the molecule as a straight chain (ignoring the 120° bond angles in the sp^2 σ system).

(a) The six *p* orbitals combine completely in phase to form the bonding orbital of lowest energy:

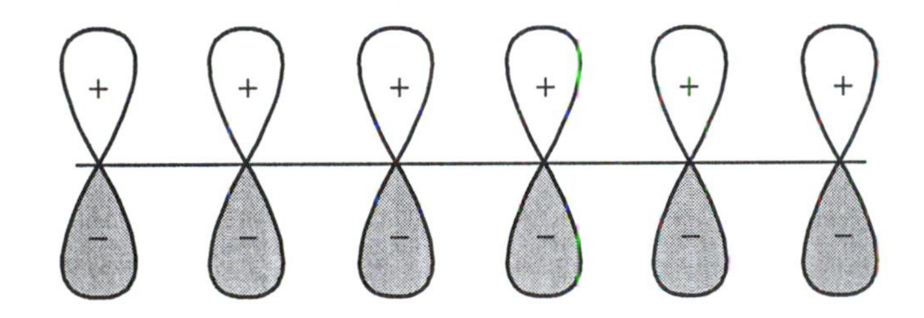

(b) The lowest π orbital contains no nodes perpendicular to the plane of the molecule.

(c) Each pair of *p* orbitals combines 180° out of phase to produce the π orbital of highest energy:

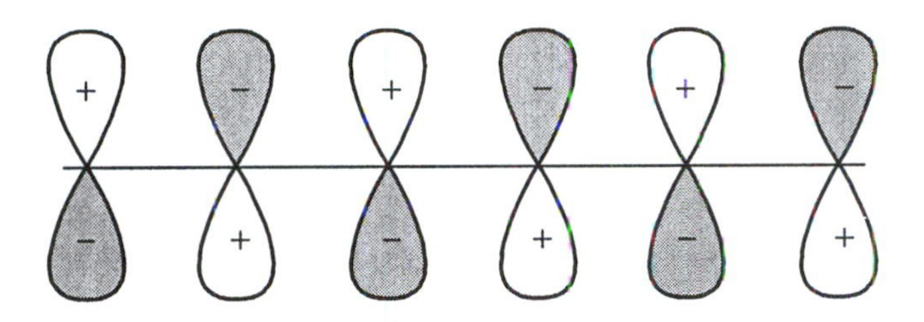

(d) The highest π orbital contains five nodes perpendicular to the molecular plane:

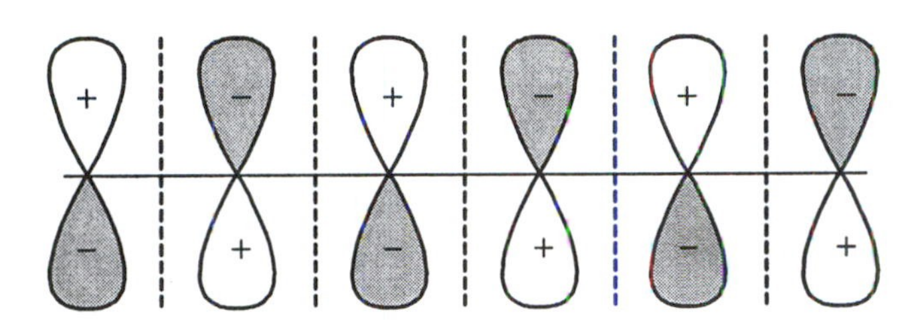

S7-19. Beryllium supplies two valence electrons, and each of the two chlorine atoms supplies seven. In all, there are 16 valence electrons to distribute.

(a) Lone pairs remain on both chlorine sites:

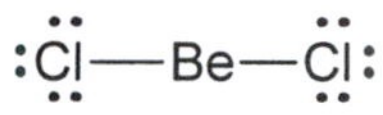

:Cl—Be—Cl:

(b) The molecule is linear. There are two bonding pairs on Be, oriented at 180° to minimize electrostatic repulsions. No nonbonding pairs are present on the central atom.

(c) The dipole moments of $BeCl_2$ and BeF_2 are the same: zero. Both molecules are linear, and the diametrically opposed bond dipoles leave the structures with no net molecular dipole moments.

S7-20. The molecule IF_5 contains 42 valence electrons, seven from each halogen atom.

(a) Six pairs of electrons—five bonding and one nonbonding—are disposed octahedrally about the central iodine site:

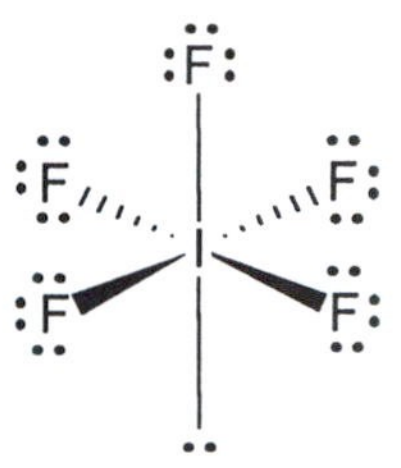

(b) The five fluorine atoms form a distorted square pyramid with iodine at the center of the base. The nonbonding electron pair, suppressed in the diagram below, points to the unoccupied sixth vertex of the octahedron:

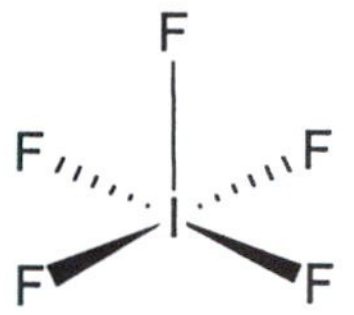

S7-21. There are 24 valence electrons to distribute: five from nitrogen, six from each oxygen, and one extra for the single negative charge.

(a) The three resonance structures differ only by successive rotations of 120°. They are all equivalent:

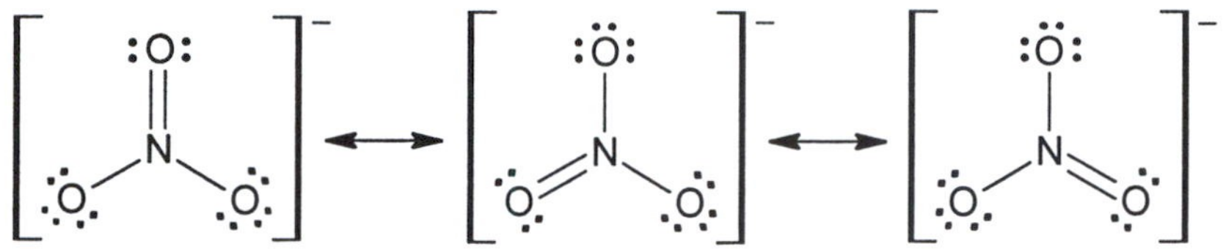

Each atom has an octet.

(b) The VSEPR model predicts a trigonal planar geometry, with bond angles of 120°. The three sets of electrons and the three atoms around nitrogen point toward the vertices of an equilateral triangle.

(c) An sp^2 hybridization of the nitrogen atom would be consistent with trigonal planar geometry.

S7-22. Eighteen valence electrons are distributed in the formate anion: four from the carbon, six from each oxygen, one from hydrogen, and one extra to yield the single negative charge.

(a) The two resonance forms below, fully equivalent, are similar to the nitrate structures considered in the preceding exercise:

(b) Like the nitrate anion, the formate structure is trigonal planar (with carbon hybridized as sp^2). Three sets of electrons surround the central atom.

(c) The π bond is delocalized over three sites: carbon and the two oxygens. See the next exercise.

S7-23. Each of the three participating atoms contributes one unhybridized p orbital to the π system.

(a) The three p orbitals combine completely in phase to produce the π molecular orbital of lowest energy:

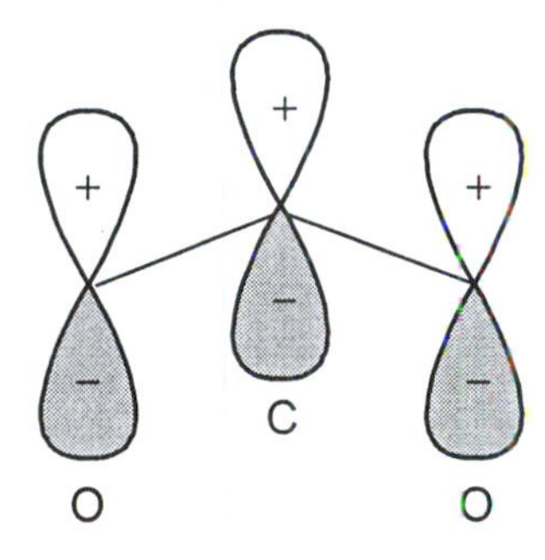

(b) Each carbon–oxygen bond consists of a full σ bond (bond order = 1) and a portion of a π bond delocalized over three sites (bond order < 1). The total bond order, inclusive of both σ and π character, is expected to fall between 1 and 2.

S7-24. See Section 7-5 in the text for a discussion of molecular orbitals in benzene.

(a) There are seven π electrons in the benzene anion (one supplied by each of the six carbons, plus one extra to yield a net negative charge). The extra electron goes into the lowest antibonding orbital:

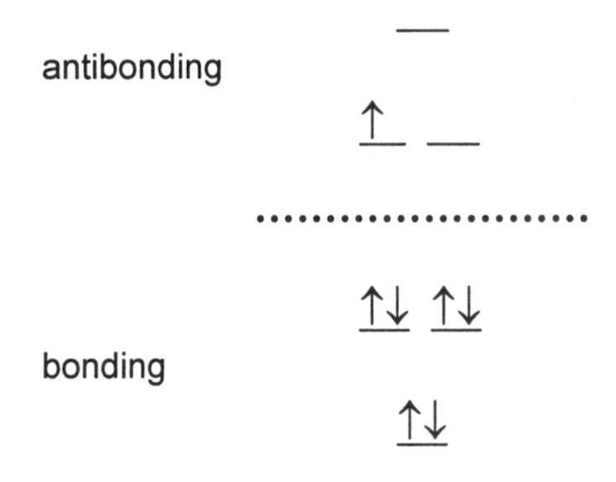

(b) The antibonding electron reduces the bond order and makes the anion less stable than the neutral molecule.

S7-25. Electron densities, interpreted as quantum mechanical probabilities, are not limited to whole numbers, simple fractions, or other such tidy forms. They derive from the square of the wave function, ψ^2, and their numerical values are generally unrestricted—provided that the total probability density adds up to the total number of electrons.

To say that we have "647 ten-thousandths of an electron" in a particular region is merely to state a probability: "647 times out of every 10,000 observations, we expect to find the electron in the location specified."

S7-26. The polarity of a bond depends on the *difference* in electronegativity between two atoms, not the absolute electronegativity of either partner alone. Combining with electropositive species such as the alkali metals, fluorine forms highly polar—typically ionic—bonds. In combination with strongly electronegative nonmetals such as oxygen or chlorine, however, the polarity is drastically reduced.

Chapter 8

Some Organic and Biochemical Species and Reactions

S8-1. Which structures may exist as open-chain alkanes?

(a) $C_{17}H_{36}$ **(b)** $C_{25}H_{50}$ **(c)** C_8H_{16} **(d)** $C_{12}H_{22}$ **(e)** $C_{11}H_{24}$

S8-2. Which structures may exist as closed rings?

(a) $C_{10}H_8$ **(b)** $C_{20}H_{42}$ **(c)** C_2H_2 **(d)** C_4H_8 **(e)** $C_6H_3Cl_3$

S8-3. Which structures may exist as planar molecules?

(a) C_2H_2BrCl **(b)** $C_2H_4Br_2$ **(c)** $C_{10}H_8$ **(d)** C_7H_{16} **(e)** C_6H_{12}

S8-4. Which structures may contain multiple bonds?

(a) $C_4H_8Br_2$ **(b)** $C_8H_{14}Cl_2$ **(c)** C_2Cl_6 **(d)** C_6H_6 **(e)** C_2H_6O

S8-5. Write a generic formula for an alkene containing n carbon atoms and m double bonds.

S8-6. Write a generic formula for an alkyne containing n carbon atoms and l triple bonds.

S8-7. Write a generic formula for a hydrocarbon containing n carbon atoms, m double bonds, and l triple bonds.

S8-8. Describe the structure and bonding of $CH_3CCCH_2CClCH_2$.

S8-9. Describe the structure and bonding of the aromatic molecule $C_6H_3Cl_3$.

S8-10. Draw any 10 structural isomers of decane, $C_{10}H_{22}$.

S8-11. How many structural isomers of $C_{18}H_{38}$ are derived from a chain of 17 carbon atoms?

S8-12. Three of the following four formulas represent exactly the same structure:

$$\begin{array}{l} H_3C{-}CH_2{-}\underset{\displaystyle CH_3}{\underset{|}{CH}}{-}CH_2{-}CH_2{-}CH_3 \end{array} \qquad \begin{array}{l} H_3C{-}CH_2{-}CH_2{-}\underset{\displaystyle CH_3}{\underset{|}{CH}}{-}CH_2{-}CH_3 \end{array}$$

$$\begin{array}{l} H_3C{-}\underset{\displaystyle \underset{\displaystyle CH_3}{\underset{|}{CH_2}}}{\underset{|}{CH}}{-}CH_2{-}CH_2{-}CH_3 \end{array} \qquad \begin{array}{l} H_3C{-}\overset{\displaystyle CH_3}{\overset{|}{\underset{\displaystyle CH_3}{\underset{|}{C}}}}{-}CH_2{-}CH_2{-}CH_3 \end{array}$$

Which one is different?

S8-13. Draw the cis and trans isomers of 2-octene.

S8-14. Draw structures for the two enantiomers of cysteine, an amino acid.

S8-15. Consider bromobenzene, C_6H_5Br, a molecule formed when one hydrogen on a benzene ring is replaced by bromine. Is bromobenzene optically active?

S8-16. Draw the ortho, meta, and para isomers of dibromobenzene.

S8-17. Suppose that each of three hydrogens on a benzene ring is replaced by group X. Draw the isomers that may result.

S8-18. Suppose that each of two hydrogens on a benzene ring is replaced by group X and that one other hydrogen is replaced by group Y. Draw the isomers that may result.

S8-19. Consider another trisubstituted benzene: one hydrogen is replaced by group X; a second hydrogen is replaced by group Y; a third is replaced by group Z. Draw the isomers that may result.

S8-20. First, redraw each molecule to show the connections explicitly. Second, identify the functional group(s):

(a) CH_3COOH **(b)** CH_3OCH_3 **(c)** CH_3CH_2Cl **(d)** $CH_3CH_2CH_2CHOHCH_3$

S8-21. First, redraw each molecule to show the connections explicitly. Second, identify the functional group(s):

(a) $HCOOH$ **(b)** $CH_3CONHCH_3$ **(c)** CH_3COCH_3 **(d)** C_2H_4

S8-22. Write a reaction that includes $CH_3CH_2CH_2COOCH_2CH_2CH_3$ as a product.

S8-23. Write a reaction in which the reactants are CH_3COOH and CH_3CH_2COOH.

S8-24. Write a reaction that describes the condensation of methionine and alanine into a dipeptide.

S8-25. Write a reaction in which the reactants are $CH_3CH_2CH_2CHCHCH_2CH_3$ and H_2.

S8-26. S_N1 reactions are generally favored in strongly polar solvents. Suggest a reason why.

SOLUTIONS

S8-1. An unbranched alkane chain of *n* carbon atoms, open at each end, contains $n - 2$ CH_2 groups in its interior and one CH_3 group at each of the two termini:

$$CH_3 \boxed{\cdots CH_2CH_2CH_2CH_2CH_2CH_2CH_2CH_2 \cdots} CH_3$$

$$(n-2)\,CH_2$$

The generic formula

$$(n-2)CH_2 + 2\,CH_3 \equiv C_nH_{2n+2}$$

is maintained in any branched isomer of the original chain, all of which derive from the same *n* carbon atoms and $2n + 2$ hydrogen atoms.

	MOLECULE	n	$2n+2$	OPEN-CHAIN ALKANE?
(a)	$C_{17}H_{36}$	17	36	yes
(b)	$C_{25}H_{50}$	25	52	no
(c)	C_8H_{16}	8	18	no
(d)	$C_{12}H_{22}$	12	26	no
(e)	$C_{11}H_{24}$	11	24	yes

S8-2. We see examples of both cycloalkanes and aromatic rings in this exercise.

(a) The molecule naphthalene, $C_{10}H_8$, consists of two fused benzene rings:

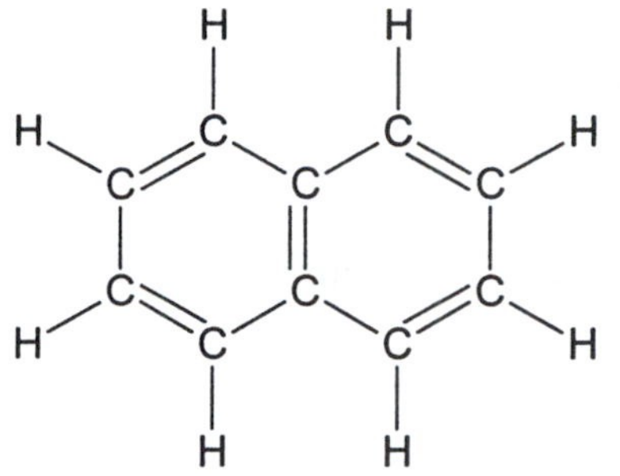

The same combination of atoms may also exist as an open-chain structure (such as, say, $HC{\equiv}C{-}C{\equiv}C{-}C{\equiv}C{-}CH{=}CH{-}CH_2CH_3$ in various isomeric forms).

(b) The molecule $C_{20}H_{42}$, conforming to the pattern C_nH_{2n+2}, is an open-chain alkane in all of its forms—not a ring. The parent structure is shown below:

$$CH_3CH_2CH_2CH_2CH_2CH_2CH_2CH_2CH_2CH_2CH_2CH_2CH_2CH_2CH_2CH_2CH_2CH_2CH_2CH_3$$

See Exercise S8-1.

(c) The molecule acetylene, C_2H_2, is linear:

$$H{-}C{\equiv}C{-}H$$

A two-carbon chain cannot close back on itself. A minimum of three carbons is needed to form a ring.

(d) Four carbons and eight hydrogens may combine to form a four-membered ring:

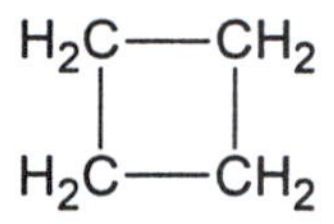

cyclobutane

In open-chain form, the molecule C_4H_8 contains a double bond:

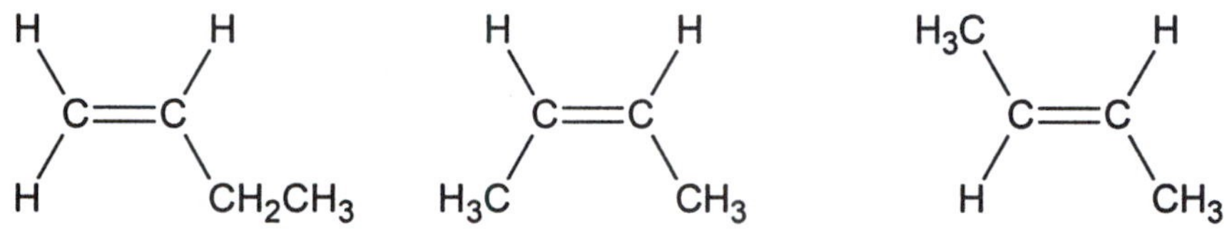

1-butene

cis-2-butene

trans-2-butene

(e) The formula $C_6H_3Cl_3$ describes a derivative of benzene:

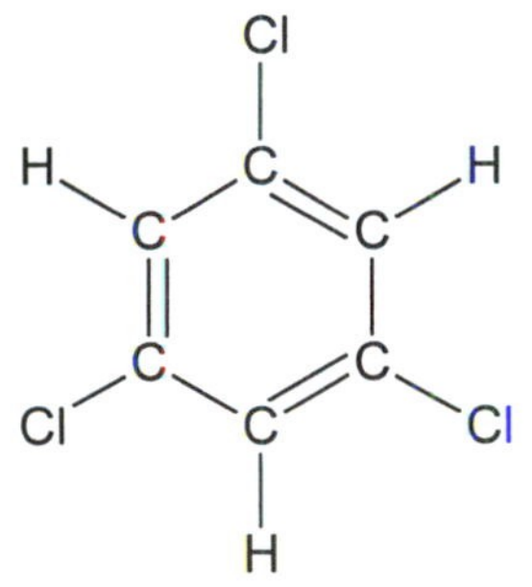

The aromatic structure exists as a closed ring. Only one isomer is shown.

Open-chain molecules are possible as well (for example, substituted derivatives of H–C≡C–CH_2–C≡C–CH_3).

S8-3. Three principal patterns dominate the geometry of hydrocarbons:

Tetrahedral (109.5°): sp^3

Trigonal planar (120°): sp^2

Linear (180°): sp

(a) The formula C_2H_2BrCl describes a derivative of ethene (ethylene):

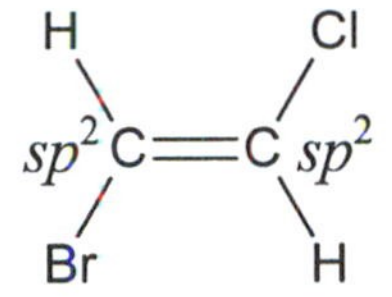

The structure, built from sp^2 carbons, is planar in all of its isomeric forms.

(b) $C_2H_4Br_2$ is a derivative of ethane, constructed from tetrahedral (sp^3) carbons:

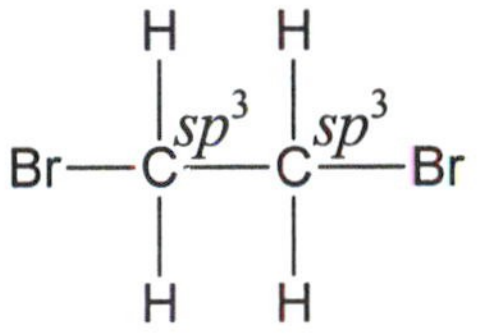

All forms of the molecule are nonplanar.

(c) See Exercise S8-2(a). Naphthalene, $C_{10}H_8$, is a planar, aromatic molecule:

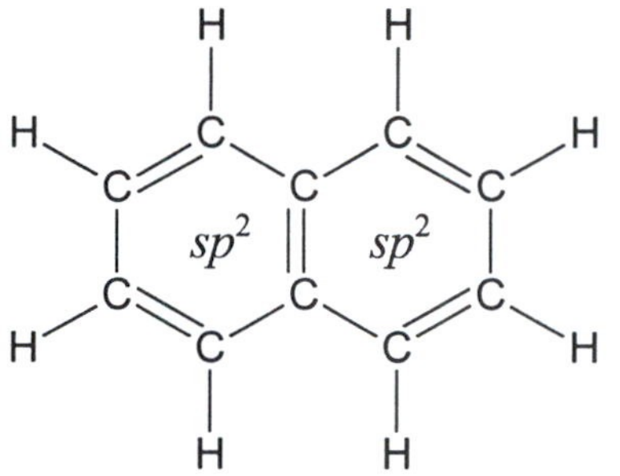

(d) C_7H_{16}, an alkane, contains seven tetrahedral (sp^3) carbons. The structure is nonplanar.

(e) The formula C_6H_{12} is consistent with a number of structures, among them an alkene (left) and a cycloalkane (right):

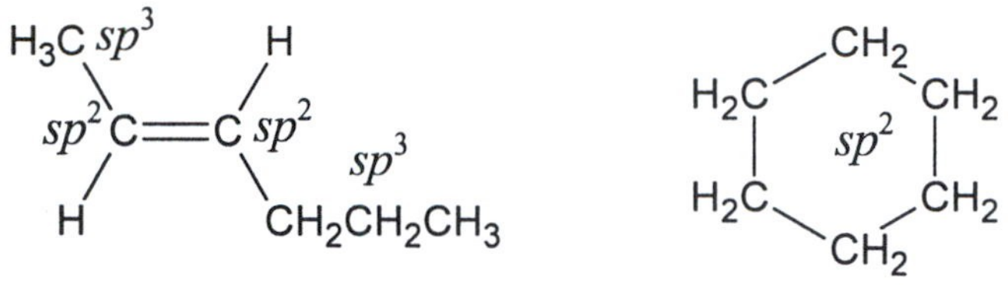

hexene (one of many isomers) cyclohexane

Neither class of molecule is planar.

S8-4. Both pure alkanes (C_nH_{2n+2}) and substituted alkanes contain only single-bonded carbon atoms.

(a) The formula $C_4H_8Br_2$ corresponds to a derivative of C_4H_{10}, an alkane. No multiple bonds are present in either the parent structure (below) or any of its isomers:

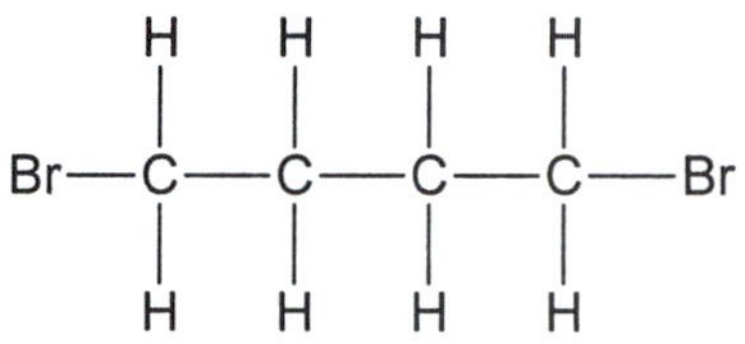

Each carbon forms four single bonds to four neighbors.

(b) $C_8H_{14}Cl_2$ is a derivative of C_8H_{16}, either an alkene or a cycloalkane. The alkene structure can support one double bond, as in the arbitrary isomer shown below:

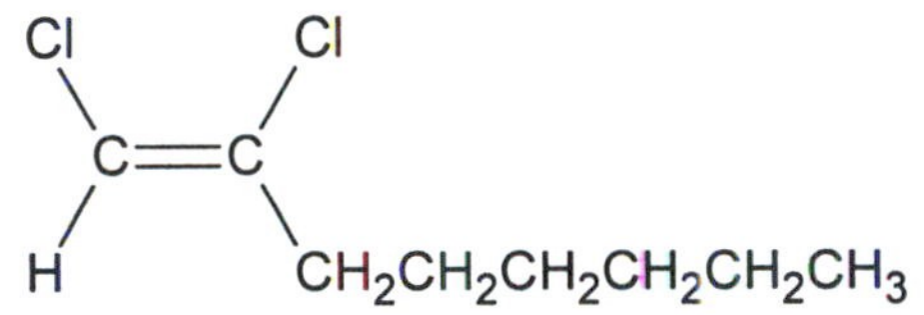

Cl Cl
C=C
H CH$_2$CH$_2$CH$_2$CH$_2$CH$_2$CH$_3$

See also Exercise S8-5.

(c) The molecule C_2Cl_6 is a chlorinated form of C_2H_6, an alkane. There are no double or triple bonds:

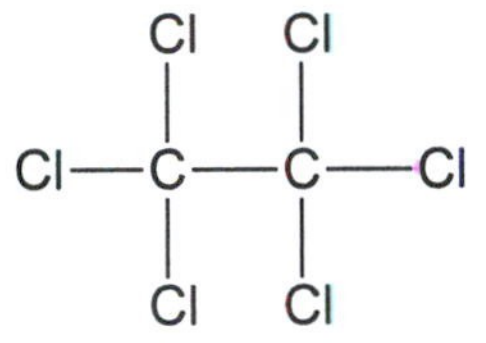

Cl Cl
Cl—C—C—Cl
Cl Cl

(d) The formula C_6H_6 does not conform to the C_nH_{2n+2} pattern of an open-chain alkane. Molecules built from six carbon atoms and six hydrogen atoms may contain both C=C and C≡C bonds, as in the two examples shown below:

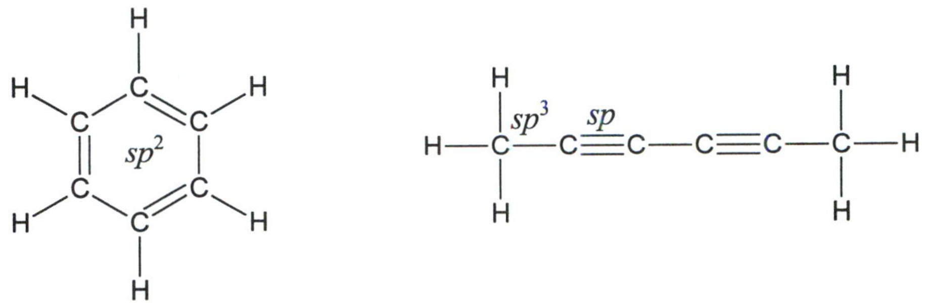

sp^2

H—C—C≡C—C≡C—C—H
sp^3 *sp*

Other multiple-bonded structures are possible as well.

(e) The formula C_2H_6O applies to derivatives of C_2H_6, an alkane. No multiple bonds are present in either of the allowed structures, illustrated below:

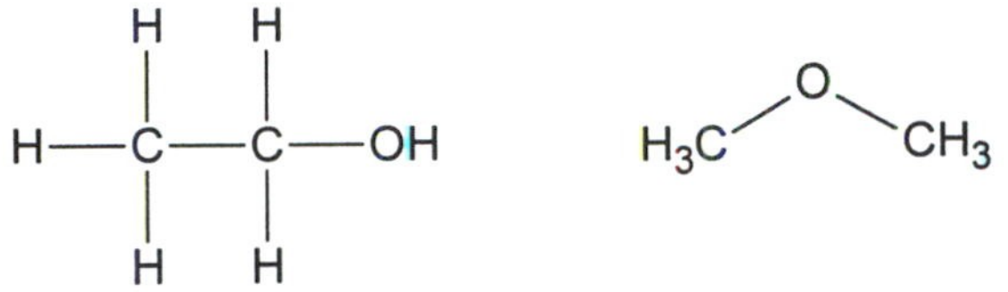

H H
H—C—C—OH
H H

H_3C O CH_3

S8-5. Start with an n-carbon alkane (C_nH_{2n+2}),

$$CH_3 \underbrace{\cdots CH_2CH_2CH_2CH_2CH_2CH_2CH_2CH_2 \cdots}_{(n-2)CH_2} CH_3$$

and subtract two hydrogen atoms for each double bond subsequently formed:

$$CH_3 \cdots CH_2CH_2CH_2{-}CH_2{-}CH_2{-}CH_2CH_2CH_2 \cdots CH_3 \;\longrightarrow\; CH_3 \cdots CH_2CH_2CH_2{-}CH{=}CH{-}CH_2CH_2CH_2 \cdots CH_3$$

The resulting alkene, stipulated to have m double bonds, contains n carbon atoms and $(2n + 2) - 2m$ hydrogen atoms. Its generic formula is $C_nH_{2(n+1-m)}$, which simplifies to C_nH_{2n} when $m = 1$.

S8-6. The approach is similar to that taken in the preceding exercise. Start with an n-carbon alkane (C_nH_{2n+2}),

$$CH_3 \underbrace{\cdots CH_2CH_2CH_2CH_2CH_2CH_2CH_2CH_2 \cdots}_{(n-2)CH_2} CH_3$$

and subtract four hydrogen atoms for each triple bond subsequently formed:

$$CH_3 \cdots CH_2CH_2CH_2{-}CH_2{-}CH_2{-}CH_2CH_2CH_2 \cdots CH_3 \;\longrightarrow\; CH_3 \cdots CH_2CH_2CH_2{-}C{\equiv}C{-}CH_2CH_2CH_2 \cdots CH_3$$

The resulting alkyne, stipulated to have l triple bonds, contains n carbon atoms and $(2n + 2) - 4l$ hydrogen atoms. Its generic formula is $C_nH_{2(n+1-2l)}$, which simplifies to C_nH_{2n-2} when $l = 1$.

S8-7. We combine the results of Exercises S8-6 and S8-7:

No. C atoms = n

$$\text{No. H atoms} = (2n+2) - \left(\frac{2}{\text{C}=\text{C}} \times m\ \text{C}=\text{C}\right) - \left(\frac{4}{\text{C}\equiv\text{C}} \times l\ \text{C}\equiv\text{C}\right) = 2(n+1-m-2l)$$

The generic formula is $C_nH_{2(n+1-m-2l)}$.

S8-8. Each carbon atom forms four bonds. Single-bonded carbon is tetrahedral (sp^3, 109.5°), double-bonded carbon is trigonal planar (sp^2, 120°), and triple-bonded carbon is linear (sp, 180°):

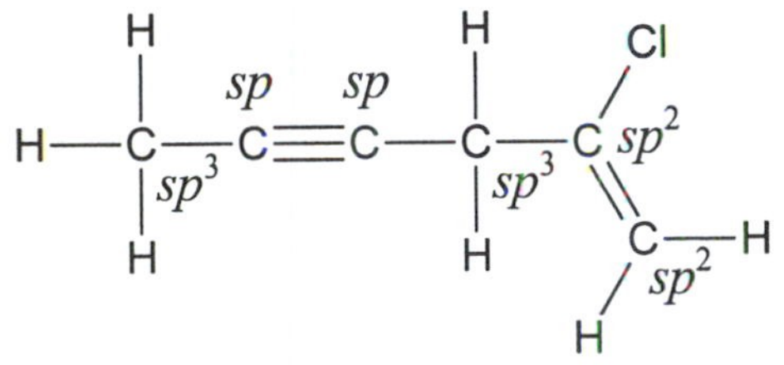

H H Cl
sp sp
H—C—C≡C—C—C sp²
sp³ sp³
H H C—H
sp²
H

The single bonds are all of σ symmetry. The double bond contains one σ linkage and one π linkage. The triple bond contains one σ linkage and two π linkages.

S8-9. The aromatic molecule $C_6H_3Cl_3$ is a derivative of benzene, represented either as the two resonance structures

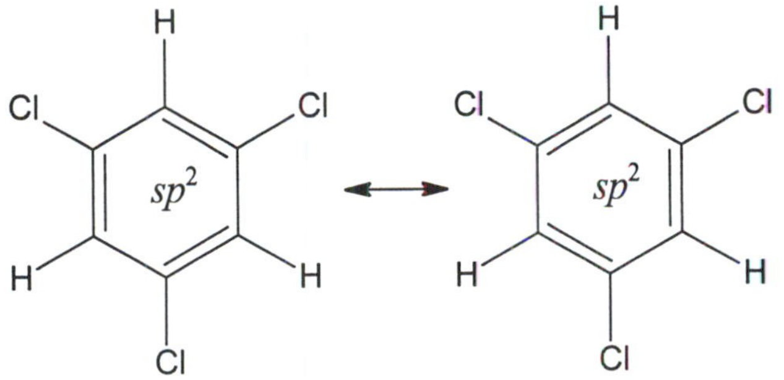

H H
Cl Cl Cl Cl
sp² ⟷ sp²
H H H H
Cl Cl

or as a delocalized form symbolized by a circle:

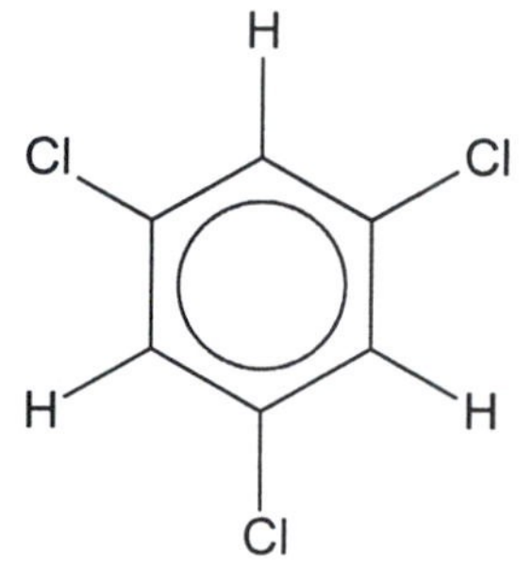

The planar carbon–carbon framework is built from a network of sp^2–sp^2 σ bonds, each originating from a trigonal carbon (120°). Six electrons are distributed throughout the delocalized π system.

For simplicity, only one isomer is depicted in the diagrams above.

S8-10. The following selection represents one possible solution. Hydrogens are omitted for simplicity.

Decane:

C—C—C—C—C—C—C—C—C—C

Derivatives of nonane:

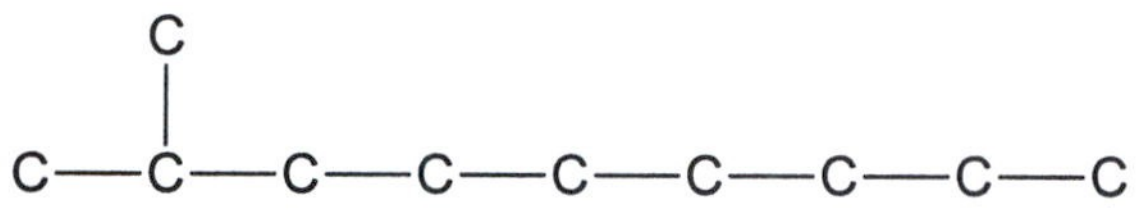

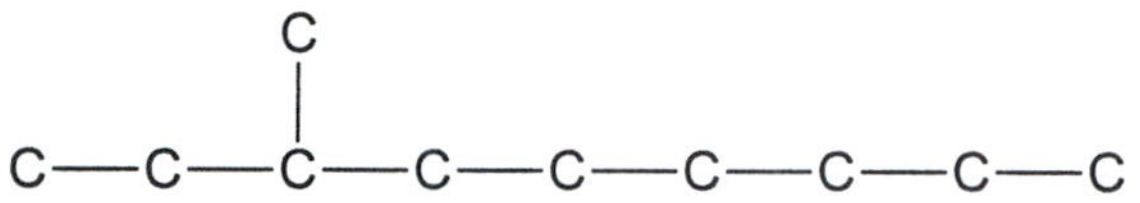

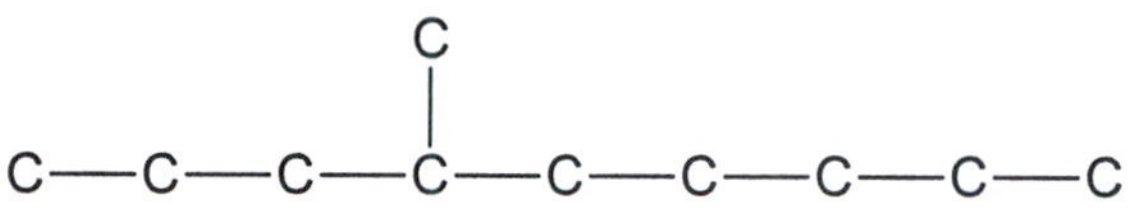

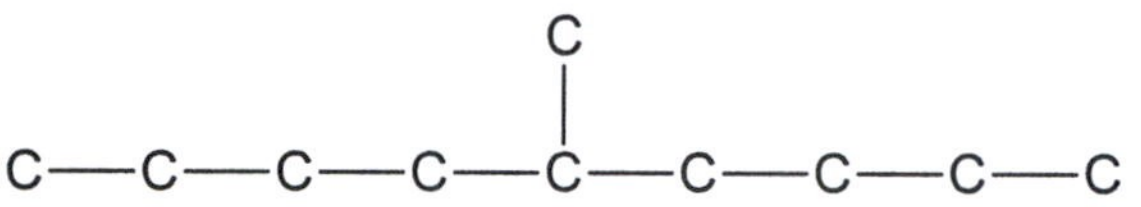

Derivatives of octane:

```
     C
     |
C—C—C—C—C—C—C—C
     |
     C
```

```
     C   C
     |   |
C—C—C—C—C—C—C—C
```

```
     C       C
     |       |
C—C—C—C—C—C—C—C
```

```
     C           C
     |           |
C—C—C—C—C—C—C—C
```

```
     C               C
     |               |
C—C—C—C—C—C—C—C
```

```
     C                   C
     |                   |
C—C—C—C—C—C—C—C
```

And so forth. Continue with derivatives of octane, distributing two methyl groups (or one ethyl group) over the chain until all nonredundant structures are covered. After that, proceed down the line with shorter chains to complete the full set of 75.

S8-11. There are eight distinguishable positions for the odd methyl group on a 17-carbon chain, as indicated below by asterisks:

```
    *   *   *   *   *   *   *   *
C—C—C—C—C—C—C—C—C—C—C—C—C—C—C—C—C
1   2   3   4   5   6   7   8   9   10  11  12  13  14  15  16  17
```

S8-12. For clarity, draw just the carbon frameworks. The three structures

```
1   2   3   4   5   6      6   5   4   3   2   1        3   4   5   6
C---C---C---C---C---C      C---C---C---C---C---C     C---C---C---C---C
        |                              |                 |
        C                              C               2 C
                                                         |
                                                       1 C
```

are equivalent renderings of the same molecule:

```
1   2   3   4   5   6
C---C---C---C---C---C
        |
        C
```

3-methylhexane

A methyl group is attached to the third carbon of hexane. The drawings differ only in the numbering and arbitrary orientation of the six-carbon chain.

The fourth structure, derived instead from a *five*-carbon chain, is different:

```
        C
1       | 2   3    4    5
C-------C-----C----C----C
        |
        C
```

2,2-dimethylpentane

Here two methyl groups are attached to the second carbon of pentane.

S8-13. The double bond runs between carbons 2 and 3 of the eight-carbon chain. In the cis isomer, the alkyl groups lie on the same side of the C=C bond. In the trans isomer, they lie on opposite sides:

```
 H         H                               H        4  5  6  7  8
  \2     3/                                 \2    3/CH2CH2CH2CH2CH3
    C===C                                     C===C
  1/     \4  5  6  7  8                     1/     \
H3C       CH2CH2CH2CH2CH3                 H3C       H
```

cis-2-octene

trans-2-octene

S8-14. Add the side chain

$$R = CH_2SH$$

to the general formula for an amino acid:

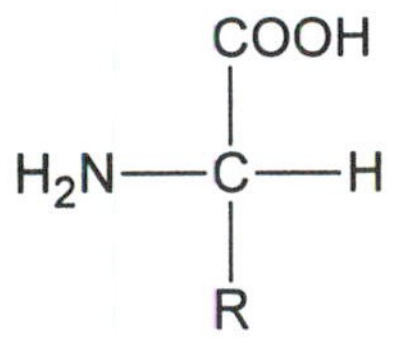

COOH
|
H_2N—C—H
|
R

The resulting molecule is *cysteine*, shown below as L and D enantiomers:

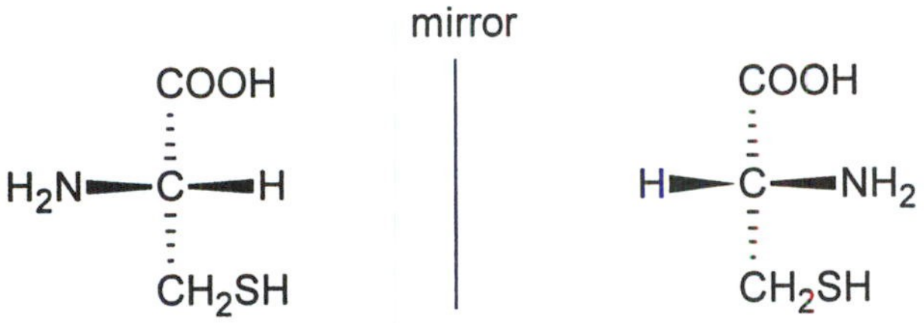

mirror

COOH | COOH
H_2N—C—H | H—C—NH_2
CH_2SH | CH_2SH
L | D

Horizontal bonds protrude out of the plane of the paper; vertical bonds retreat into the plane.

S8-15. Bromobenzene is not optically active. The planar molecule is identical to its mirror image:

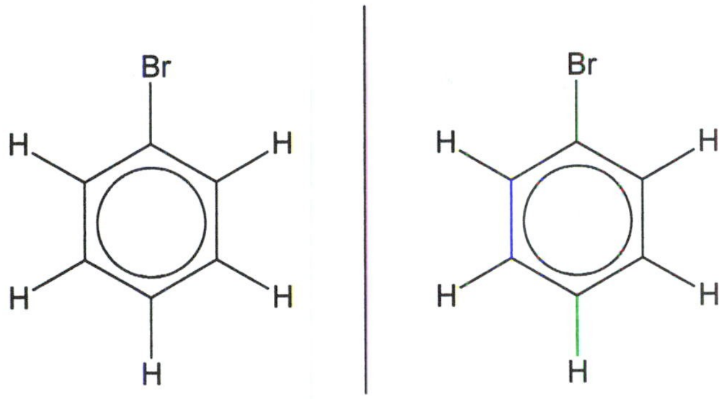

Br Br
H H H H
H H H H
H H

S8-16. Substituents in the ortho, meta, and para isomers appear at positions 1-2, 1-3, and 1-4, respectively:

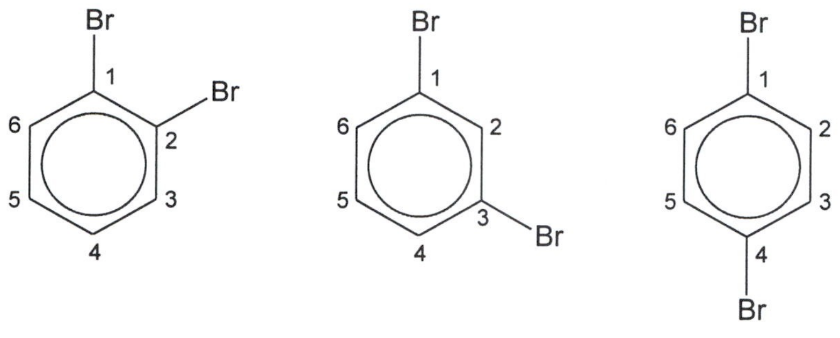

ortho meta para

Hydrogens are omitted for clarity.

S8-17. There are three nonredundant isomers of a homogeneously trisubstituted benzene:

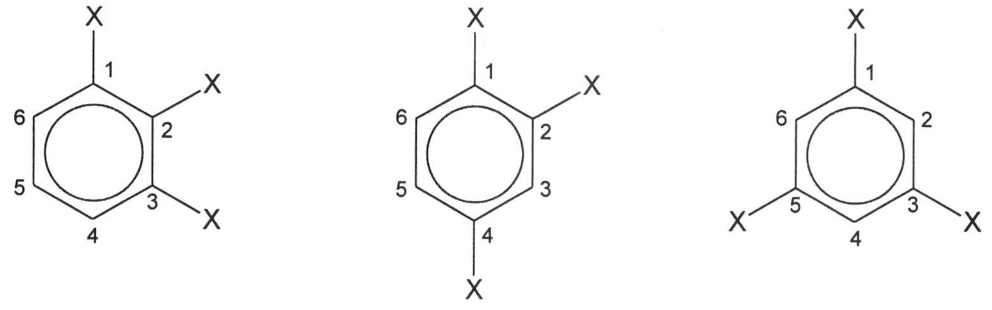

1,2,3 1,2,4 1,3,5

All other forms are related by rotations and are therefore equivalent. For example:

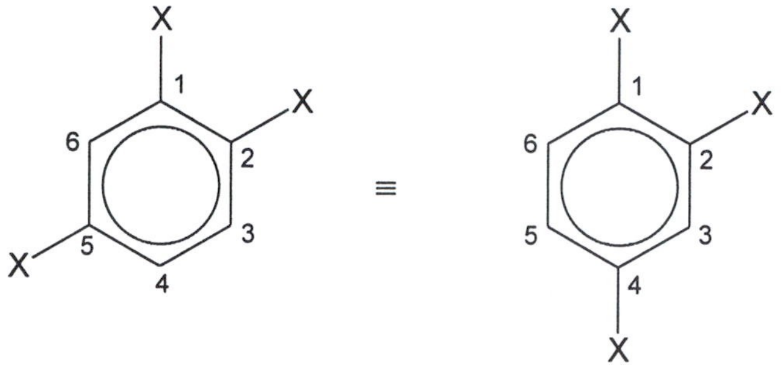

1,2,5 ≡ 1,2,4

S8-18. There are six isomers. See the display at the top of the next page.

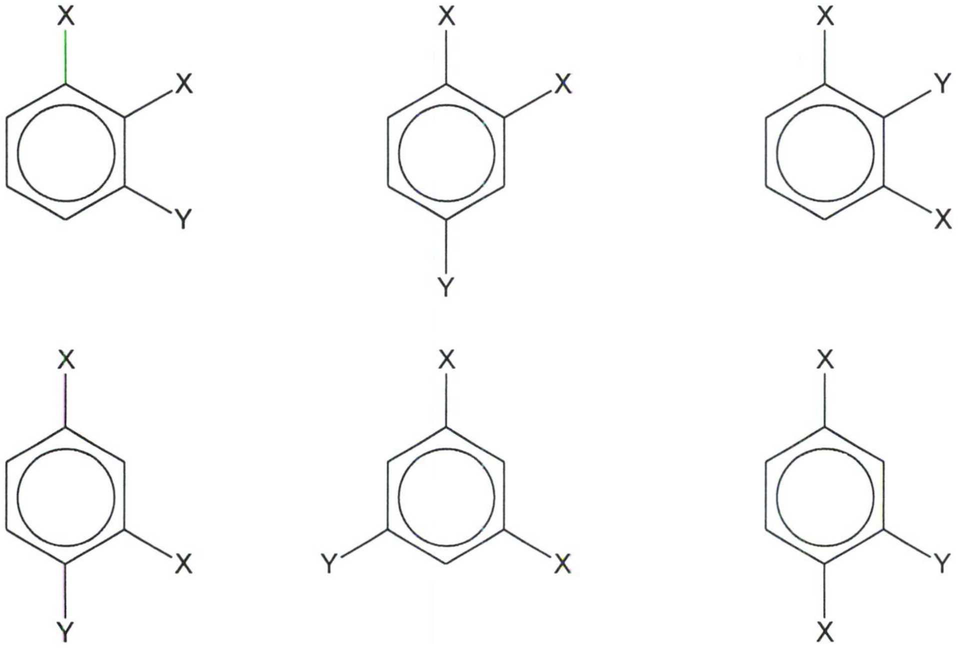

X X Y X X Y X Y X
Y X X X X Y Y
Y X

S8-19. There are ten isomers:

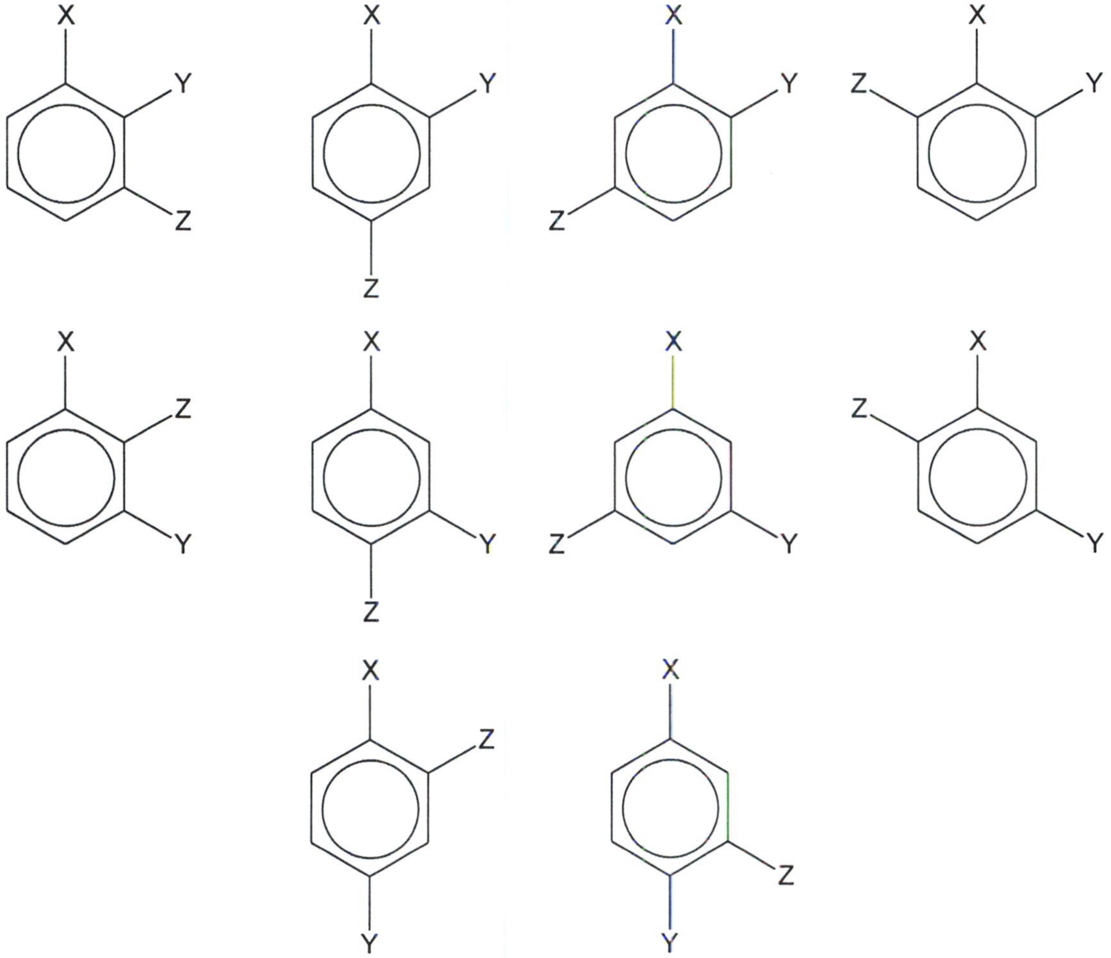

X X X X
Y Y Y Z Y
Z Z Z
X X X X
Z Z
Y Y Z Y Y
Z
X X
Z
Z
Y Y

S8-20. Each carbon atom makes four bonds. Angles around single-bonded (sp^3) carbons are understood to be approximately 109.5°.

(a) Carboxylic acid:

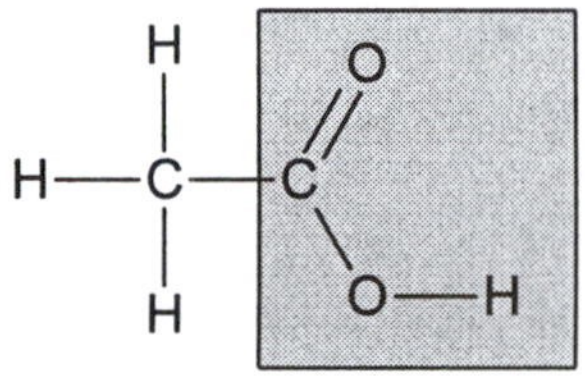

(b) Ether:

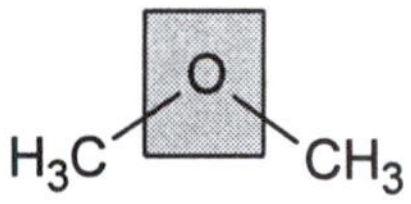

(c) Alkyl halide:

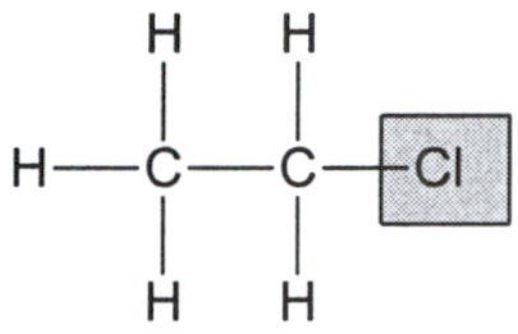

(d) Alcohol:

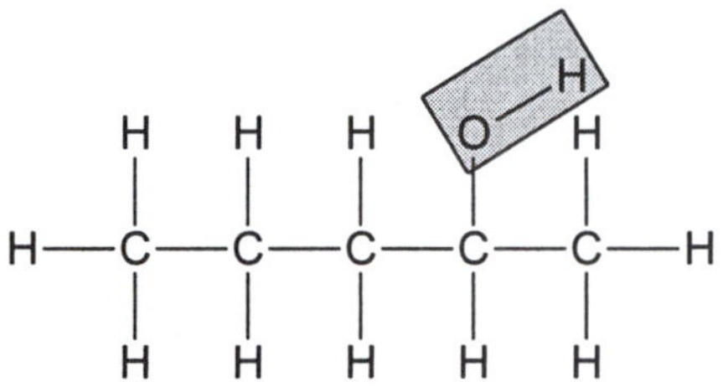

S8-21. Similar.

(a) Carboxylic acid:

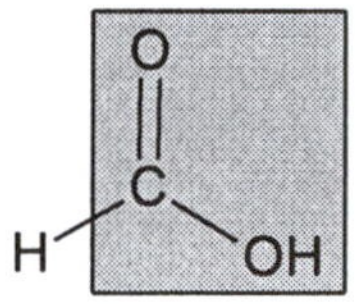

(b) Amide:

$$H_3C-C(=O)-N(H)-CH_3$$

(c) Ketone:

$$H_3C-C(=O)-CH_3$$

(d) Alkene:

$$H_2C=CH_2$$

S8-22. The ester can be produced by condensation of a carboxylic acid and an alcohol:

$$CH_3CH_2CH_2-C(=O)-OH + HO-CH_2CH_2CH_3 \longrightarrow CH_3CH_2CH_2-C(=O)-OCH_2CH_2CH_3 + H_2O$$

S8-23. Condensation of two carboxylic acids will produce an anhydride:

$$H_3C-C(=O)-OH + HO-C(=O)-CH_2CH_3 \longrightarrow H_3C-C(=O)-O-C(=O)-CH_2CH_3 + H_2O$$

S8-24. Each amino acid may be written in the form

$$\mathrm{H_2N{-}CH(R){-}C({=}O){-}OH}$$

Formation of a dipeptide is brought about by a condensation reaction:

$$\mathrm{H_2N{-}CH(R_{Met}){-}C({=}O){-}OH} + \mathrm{H{-}NH{-}CH(R_{Ala}){-}COOH} \longrightarrow \mathrm{H_2N{-}CH(R_{Met}){-}C({=}O){-}NH{-}CH(R_{Ala}){-}COOH} + H_2O$$

Carbons hybridized as sp^3 are understood to be configured tetrahedrally. Specific amino-acid side chains are noted below:

$$R_{Met} = CH_2CH_2SCH_3$$

$$R_{Ala} = CH_3$$

S8-25. Addition of molecular hydrogen to the alkene produces the corresponding alkane:

$$\mathrm{H_7C_3(H)C{=}C(H)C_2H_5} + H_2 \longrightarrow \mathrm{H_7C_3{-}CH_2{-}CH_2{-}C_2H_5}$$

S8-26. A polar solvent can help stabilize the transient carbocation that develops during the course of an S_N1 reaction.

Chapter 9

States of Matter

S9-1. Temperature is a measure of the average kinetic energy of particles in motion. Would you expect the volume of a gas to increase, decrease, or remain the same as the temperature is increased? Assume that the gas remains under constant atmospheric pressure throughout.

S9-2. Would you expect the volume of a gas to increase, decrease, or remain the same as the system is subjected to increasing external pressure? Assume that the temperature remains constant.

S9-3. Under what conditions of temperature and pressure would you expect a gas to condense into a liquid?

S9-4. Under what conditions of pressure and volume would you expect a gas to condense into a liquid?

S9-5. **(a)** Estimate, to within a power of 10, the number of molecules contained in the air filling an average-sized bedroom. **(b)** Estimate the corresponding total mass of air. In each case, state the assumptions you make in order to arrive at the numbers you do. Turn to whatever sources of information you deem helpful.

S9-6. Which substance in each pair is more likely to exist as a gas at room temperature?

(a) H_2O, C_2H_6 **(b)** Rn, Li **(c)** C_3H_8, C_6H_{14} **(d)** NaCl, HCl

S9-7. What are the principal interactions holding together each of the following phases?

(a) Mg(s) **(b)** $Cl_2(g)$ **(c)** $MgCl_2(s)$ **(d)** $MgCl_2(aq)$

S9-8. What are the principal interactions holding together each of the following phases?

(a) $H_2(g)$ **(b)** $Cl_2(s)$ **(c)** $HCl(\ell)$ **(d)** $HCl(aq)$

S9-9. Describe the interactions responsible for maintaining the primary, secondary, and tertiary structures of a protein.

S9-10. Which substance in each pair is more likely to boil at the higher temperature?

(a) $CH_3CH_2CH_2CH_2CH_3$, $CH_3CH_2CH_2CH_3$
(b) $CH_3CH_2CH_2CH_2CH_2OH$, $CH_2OHCHOHCH_2CH_2CH_2OH$
(c) $C_2H_4Cl_2$, C_2H_6
(d) CH_3OCH_3, H_2O

S9-11. Which substance in each pair is more likely to boil at the higher temperature?

(a) Xe, Kr **(b)** CH_3OH, CF_4 **(c)** C_3H_8, C_3F_8 **(d)** C_2F_6, C_2H_5F

S9-12. **(a)** Will a gas consisting of He atoms conduct electricity? **(b)** The term *plasma* refers to an electrically neutral gas consisting of electrons and ions. Will a helium plasma conduct electricity? **(c)** Under what conditions of temperature will a plasma form?

S9-13. **(a)** In what way is a plasma (see the preceding exercise) similar to a metal? How is it different? **(b)** In what way is a plasma similar to an electrolytic solution? In what way is it different?

S9-14. Which substance in each pair is more likely to have the higher lattice energy?

(a) $CaCl_2$, KCl **(b)** NaCl, CsCl **(c)** $SrCl_2$, $MgCl_2$

S9-15. Gold (Au) crystallizes in a cubic lattice with cell dimension equal to 4.079 Å and with density equal to 19.282 g cm^{-3}. Is the unit cell a face-centered, body-centered, or simple cubic structure?

S9-16. How many atoms of metallic silver are found in a spherical sample with radius equal to 0.107 m? Consult Appendix C for any necessary data.

S9-17. How many molecules of H_2O are found in a spherical sample with radius equal to 0.107 m? Consult Appendix C for any necessary data.

S9-18. Assume that 1.00 mol He occupies a volume of 22.4 L. **(a)** How many atoms of He are found in a spherical sample with radius equal to 0.107 m? **(b)** Compare the value obtained here with the values obtained in the preceding two exercises.

S9-19. How many grams of $MgCl_2$ must be dissolved in a total volume of 365 mL to produce a chloride concentration of 0.00763 *M*?

S9-20. A 50.0-mL solution of glucose ($C_6H_{12}O_6$) has a concentration of 0.436 *M*. To what volume must the solution be diluted in order to make the concentration 0.200 *M*?

S9-21. A mixture of gases contains 5.76 g He, 50.6 g N_2, and 84.4 g CO_2. Calculate the mole fraction of each component.

S9-22. A mixture of ethanol and methanol contains 1.76 g CH_3OH, equivalent to a mole fraction of 0.287. Calculate the mass of ethanol present.

S9-23. Suppose that the mole fraction of ethanol is 0.459 in a certain aqueous solution. Calculate the ratio—by mass—of ethanol to water in the solution.

SOLUTIONS

S9-1. Molecules made to move *faster*—with greater kinetic energy—will spread into a larger volume, provided that the gas is free to displace its confining walls.

Consider: Colliding against the walls with greater velocity and momentum, particles at a higher temperature temporarily generate a higher pressure inside the container. This overpressure is dissipated as the gas expands and the system comes into balance once again with its surroundings. The expansion stops when the internal pressure of the gas is equal to the external pressure of the atmosphere, whereupon there is no further tendency for either the gas to push out or the surroundings to push in.

We anticipate, then, that the volume of a gas will increase with temperature if the system is allowed to expand under constant pressure. This statement provides a summary of Charles's law, to be discussed further in Chapter 10.

S9-2. A gas squeezed under a greater external pressure will shrink into a smaller volume, so long as the average kinetic energy of the particles remains unchanged.

Our argument runs parallel to the one we advanced in Exercise S9-1: If the gas and its surroundings are to remain in balance, then any increase in external pressure demands a reciprocating increase in *internal* pressure. As the container becomes smaller, the particles collide more frequently with the walls and consequently deliver a higher pressure. The average velocity, momentum, and kinetic energy stay constant throughout, since the temperature is fixed.

Thus we anticipate Boyle's law, as it will be presented in Chapter 10: Volume is inversely proportional to pressure at constant temperature.

S9-3. Condensation will occur when the particles are able to interact most effectively—when they are moving slowly (at low temperature) and when they are compressed into a small volume (under high pressure).

S9-4. Similar. Condensation will take place under conditions that crowd the particles into close quarters, thereby increasing the likelihood of intermolecular interactions: high pressure, low volume.

S9-5. This exercise anticipates the exposition of Avogadro's law in Chapter 10, which states:

> Equal volumes of gases at the same temperature and pressure contain equal numbers of particles.

The argument is plausible, even at the level considered here in Chapter 9. So long as intermolecular interactions remain unexpressed in a gas (any gas), we expect the mechanical properties to depend very little on the chemical identity of the particles. All gases should behave universally, at least to some extent. See Section 9-4 in *Principles of Chemistry*.

(a) We already know, from Example 9-1 in the text, that one mole of an atmospheric gas such as oxygen occupies 22.4 L at standard temperature and pressure. Avogadro's number subsequently enables us to convert moles per unit volume into molecules per unit volume.

To obtain a specific number of molecules, we shall assume dimensions of 3.0 m × 4.5 m × 2.5 m (approximately 10 ft × 15 ft × 8 ft):

$$\left(\frac{1 \text{ mol}}{22{,}400 \text{ mL}} \times \frac{1 \text{ mL}}{\text{cm}^3} \times \frac{6.02 \times 10^{23} \text{ particles}}{\text{mol}}\right)\left(300 \text{ cm} \times 450 \text{ cm} \times 250 \text{ cm}\right)$$

$$= 9.1 \times 10^{26} \text{ particles}$$

(b) Here we assume an average molar mass of approximately 29 g, in rough agreement with the 4:1 proportions of N_2 and O_2 in the atmosphere (see text, page 352):

$$\left(\frac{1 \text{ mol}}{22{,}400 \text{ mL}} \times \frac{1 \text{ mL}}{\text{cm}^3} \times \frac{29 \text{ g}}{\text{mol}}\right)\left(300 \text{ cm} \times 450 \text{ cm} \times 250 \text{ cm}\right)$$

$$= 4.4 \times 10^4 \text{ g}$$

S9-6. The system with the weaker intermolecular interactions is more likely to exist as a gas at room temperature.

(a) The nonpolar molecule ethane, C_2H_6, interacts only by way of the weak London dispersion force, whereas H_2O is able to form hydrogen bonds (much stronger). Ethane is a gas at room temperature; water is a liquid.

(b) Rn interacts via the London dispersion force. Li forms metallic bonds. Atoms of radon, a noble gas, thus are interconnected far more weakly than atoms of lithium, a solid at room temperature.

(c) Both C_3H_8 and C_6H_{14} have access only to the weak London force, but the more massive C_6H_{14} interacts more strongly. Propane (C_3H_8) is more likely to be a gas.

(d) NaCl is an ionic solid, held together by the strongest of the interatomic and intermolecular forces: ion-ion Coulomb interactions. HCl, by contrast, is a covalent molecule that interacts by way of dipole–dipole forces, much weaker than ionic bonds. Hydrogen chloride is more likely to be a gas.

See also the comment to Exercise S9-8(c) below.

S9-7. Additional practice in classifying noncovalent interactions.

	PHASE	ACTORS	INTERACTION
(a)	Mg(s)	valence electrons, lattice cations	metallic bonds
(b)	Cl_2(g)	nonpolar molecules	London dispersion force
(c)	$MgCl_2$(s)	cations, anions	ionic bonds (Coulomb force)
(d)	$MgCl_2$(aq)	cations, anions, polar molecules	ion–dipole coupling

S9-8. Similar.

	PHASE	ACTORS	INTERACTION
(a)	H_2(g)	nonpolar molecules	London dispersion force
(b)	Cl_2(s)	nonpolar molecules	London dispersion force
(c)	HCl(ℓ)	polar molecules	dipole–dipole coupling*
(d)	HCl(aq)	H_3O^+, Cl^-, H_2O	ion–dipole coupling, H bonds

*Note that molecules of HCl are also able to form hydrogen bonds, although the linkages are not as strong as those involving N, O, or F.

S9-9. The primary structure is maintained by covalent amide linkages between adjacent amino acids (peptide bonds), as described in Section 8-4 of *Principles of Chemistry*. The secondary structure arises from noncovalent interactions (largely hydrogen bonds) between amide C=O and N–H groups on nearby amino acids. The tertiary structure arises from noncovalent interactions between distant amino acids, including hydrogen bonds, ionic bonds, dipole–dipole couplings, and the London dispersion force. Covalent S–S linkages are also possible.

S9-10. The system with the stronger intermolecular interactions should boil at the higher temperature.

(a) $CH_3CH_2CH_2CH_2CH_3$, more massive than $CH_3CH_2CH_2CH_3$, experiences the stronger London interactions and thus has the higher boiling point.

(b) $CH_2OHCHOHCH_2CH_2CH_2OH$, able to form more OH hydrogen bonds and also more massive than $CH_3CH_2CH_2CH_2CH_2OH$, boils at the higher temperature.

(c) Even without knowing whether or not $C_2H_4Cl_2$ is polar, we do know that it is heavier than C_2H_6—and hence subject to stronger dispersion interactions. The chlorinated molecule has the higher boiling point.

(d) H_2O, able to form hydrogen bonds, boils at a higher temperature than the moderately polar molecule CH_3OCH_3.

S9-11. The reasoning is the same as in the preceding exercise.

(a) The heavier atom—xenon—enjoys stronger dispersion interactions and consequently has a higher boiling point than krypton.

(b) CH_3OH forms hydrogen bonds, whereas nonpolar CF_4 interacts only by means of the dispersion interaction. Methanol (CH_3OH) boils at the higher temperature.

(c) The more massive molecule—C_3F_8—has the higher boiling point.

(d) C_2H_5F, endowed with a dipole moment, has a higher boiling point than nonpolar C_2F_6.

S9-12. We make mention of the *plasma* phase, sometimes called a "fourth state of matter."

(a) A gas of helium atoms will not conduct electricity. Each atom keeps its electrons and remains a neutral particle, precluding any flow of current.

(b) With electrons stripped away from the atoms, a plasma of electrons and helium ions will conduct electricity. Charged particles are able to flow freely.

(c) Exceedingly high kinetic energies (such as realized at high temperature) are required to ionize the atoms and produce a plasma.

S9-13. Plasma, continued.

(a) Both a metal and a plasma contain charged particles able to move freely: the valence electrons of a metal, and the free electrons and cations of a plasma. In a metal, however, the cations usually occupy fixed positions in a solid-state lattice.

(b) Again, both phases contain mobile charged particles and thus are able to conduct electricity. Furthermore—like a plasma (but unlike most metals)—the positive ions in an electrolytic solution are free to move as well.

The two phases differ in the identity of their charge carriers. The negative particles in a plasma are free electrons, whereas the negative particles in an electrolytic solution are atomic or molecular ions. In addition, the solution exists as a condensed phase. The plasma usually does not.

S9-14. The Coulomb force between charges q_1 and q_2 is proportional to q_1q_2/r^2, where r is the distance separating the charges. The corresponding potential energy varies as $1/r$.

Bear in mind, below, that atomic and ionic radii of main-group elements typically increase going down a column and decrease going across a row. Cesium is larger than sodium. Strontium is larger than magnesium. Potassium is larger than calcium.

(a) Doubly positive Ca^{2+} binds chlorine more strongly than singly positive K^+, and the smaller calcium ion permits a closer packing and stronger potential as well. $CaCl_2$ has a larger lattice energy than KCl.

(b) Na^+, an ion smaller than Cs^+, produces a stronger lattice in combination with the chloride anion. The lattice energy of NaCl is higher than that of CsCl.

(c) Similarly, $MgCl_2$ has a higher lattice energy than $SrCl_2$. Mg^{2+} is the smaller of the two cations.

S9-15. Use the density and volume

$$\rho = 19.282 \text{ g cm}^{-3}$$

$$V = a^3 = \left(4.079 \times 10^{-10} \text{ m}\right)^3$$

to calculate the number of atoms in each cubic unit cell:

$$\frac{19.282 \text{ g}}{\text{cm}^3} \times \left(\frac{100 \text{ cm}}{\text{m}}\right)^3 \times \frac{1 \text{ mol Au}}{196.96655 \text{ g}} \times \frac{6.0221 \times 10^{23} \text{ atoms}}{\text{mol}} \times \left(4.079 \times 10^{-10} \text{ m}\right)^3$$

$$= 4.001 \text{ atoms Au}$$

The result, four atoms, is consistent with a face-centered structure.

S9-16. From Table C-9 in *Principles of Chemistry* we obtain the density of metallic silver:

$$\rho = 10.50 \text{ g mL}^{-1} = 10.50 \text{ g cm}^{-3}$$

Combining this value with the volume of a cube,

$$V = \tfrac{4}{3}\pi r^3 \qquad \text{(radius} = r\text{)}$$

we then calculate the number of atoms:

$$\frac{10.50 \text{ g}}{\text{cm}^3} \times \left(\frac{100 \text{ cm}}{\text{m}}\right)^3 \times \frac{1 \text{ mol Ag}}{107.8682 \text{ g}} \times \frac{6.0221 \times 10^{23} \text{ atoms}}{\text{mol}} \times \frac{4\pi}{3}(0.107 \text{ m})^3$$

$$= 3.01 \times 10^{26} \text{ atoms Ag}$$

S9-17. Assume that the density of water is 1.00 g cm^{-3}, and use the formula $V = \tfrac{4}{3}\pi r^3$ once again to calculate the number of molecules:

$$\frac{1.00 \text{ g}}{\text{cm}^3} \times \left(\frac{100 \text{ cm}}{\text{m}}\right)^3 \times \frac{1 \text{ mol H}_2\text{O}}{18.015 \text{ g}} \times \frac{6.0221 \times 10^{23} \text{ molecules}}{\text{mol}} \times \frac{4\pi}{3}(0.107 \text{ m})^3$$

$$= 1.72 \times 10^{26} \text{ molecules H}_2\text{O}$$

The value for water, a liquid, is roughly the same as for metallic silver, a solid.

S9-18. Here we demonstrate the large difference in density between gaseous and condensed phases.

(a) Take the same approach as in Exercises S9-16 and S9-17, this time using a molar volume of 22.4 L (equivalently, 22,400 cm^3):

$$\frac{1.00 \text{ mol He}}{22{,}400 \text{ cm}^3} \times \left(\frac{100 \text{ cm}}{\text{m}}\right)^3 \times \frac{6.0221 \times 10^{23} \text{ atoms}}{\text{mol}} \times \frac{4\pi}{3}(0.107 \text{ m})^3 = 1.38 \times 10^{23} \text{ atoms He}$$

(b) The density of the gas is approximately 1000 times smaller than the density of either the liquid or solid. Correspondingly fewer particles are distributed over the same volume.

S9-19. Dissolution of one mole of $MgCl_2$ yields two moles of Cl^- ions:

$$\left(\frac{0.00763 \text{ mol Cl}^-}{\text{L}} \times \frac{1 \text{ L}}{1000 \text{ mL}} \times 365 \text{ mL}\right) \times \frac{1 \text{ mol MgCl}_2}{2 \text{ mol Cl}^-} \times \frac{95.211 \text{ g MgCl}_2}{\text{mol MgCl}_2}$$

$$= 0.133 \text{ g MgCl}_2$$

S9-20. In general, problems of this type can always be solved by application of a simple conservation equation:

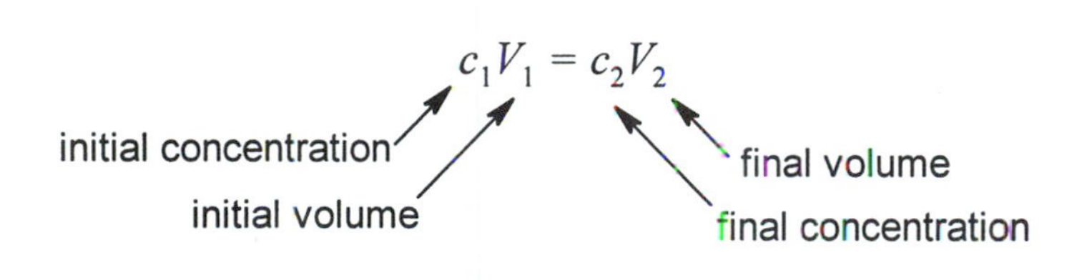

The governing relationship

$$\text{Concentration} \times \text{volume} = \text{amount}$$

$$\frac{\text{mol}}{\text{L}} \times \text{L} = \text{mol}$$

thus affirms that the amount of solute remains constant before and after dilution.

Here, in our specific example, we have a fixed 0.0218 mole of solute in the initial and final solutions:

$$\begin{array}{ccc} \text{CONCENTRATION} & \text{VOLUME} & \text{AMOUNT} \end{array}$$

$$\left(\frac{0.436 \text{ mol}}{\text{L}} \times \frac{1 \text{ L}}{1000 \text{ mL}}\right)(50.0 \text{ mL}) = 0.0218 \text{ mol}$$

The new volume follows directly:

$$c_1V_1 = c_2V_2$$

$$V_2 = \frac{c_1}{c_2}V_1 = \frac{0.436\ M}{0.200\ M} \times 50.0 \text{ mL} = 109 \text{ mL}$$

S9-21. Calculate first the amount of each component,

$$n_{\text{He}} = 5.76 \text{ g} \times \frac{1 \text{ mol}}{4.003 \text{ g}} = 1.439 \text{ mol}$$

$$n_{\text{N}_2} = 50.6 \text{ g} \times \frac{1 \text{ mol}}{28.013 \text{ g}} = 1.806 \text{ mol}$$

$$n_{\text{CO}_2} = 84.4 \text{ g} \times \frac{1 \text{ mol}}{44.010 \text{ g}} = 1.918 \text{ mol}$$

and then apply the definition of mole fraction:

$$X_{He} = \frac{n_{He}}{n_{He} + n_{N_2} + n_{CO_2}} = \frac{1.439 \text{ mol}}{1.439 \text{ mol} + 1.806 \text{ mol} + 1.918 \text{ mol}} = 0.279$$

$$X_{N_2} = \frac{n_{N_2}}{n_{He} + n_{N_2} + n_{CO_2}} = \frac{1.806 \text{ mol}}{1.439 \text{ mol} + 1.806 \text{ mol} + 1.918 \text{ mol}} = 0.350$$

$$X_{CO_2} = \frac{n_{CO_2}}{n_{He} + n_{N_2} + n_{CO_2}} = \frac{1.918 \text{ mol}}{1.439 \text{ mol} + 1.806 \text{ mol} + 1.918 \text{ mol}} = 0.371$$

To check, verify that the sum of all the mole fractions is equal to 1:

$$X_{He} + X_{N_2} + X_{CO_2} = 0.279 + 0.350 + 0.371 = 1.000$$

S9-22. Start with the mole fraction of component 2 in a two-component mixture,

$$X_2 = \frac{n_2}{n_1 + n_2}$$

and solve for the molar amount of component 1:

$$n_2 = n_1 X_2 + n_2 X_2$$

$$n_1 X_2 = n_2 - n_2 X_2$$

$$n_1 = \frac{n_2(1 - X_2)}{X_2}$$

Then use the mass and mole fraction of CH_3OH (component 2) to calculate the molar and gram amounts of CH_3CH_2OH (component 1):

$$n_1 = \frac{\left(1.76 \text{ g CH}_3\text{OH} \times \dfrac{1 \text{ mol}}{32.042 \text{ g CH}_3\text{OH}}\right)(1 - 0.287)}{0.287} \times \frac{46.069 \text{ g CH}_3\text{CH}_2\text{OH}}{\text{mol}}$$

$$= 6.29 \text{ g CH}_3\text{CH}_2\text{OH}$$

S9-23. First, define the following symbols:

n_1 = moles of H_2O $\qquad$ X_1 = mole fraction of H_2O

n_2 = moles of CH_3CH_2OH $\qquad$ X_2 = mole fraction of CH_3CH_2OH

Next, use the relationship

$$X_1 + X_2 = 1$$

to express X_1 in terms of X_2:

$$X_1 = 1 - X_2$$

Calculate, after that, the ratio of moles from the ratio of mole *fractions*,

$$\frac{X_2}{X_1} = \frac{\dfrac{n_2}{n_1 + n_2}}{\dfrac{n_1}{n_1 + n_2}} = \frac{n_2}{n_1} = \frac{X_2}{1 - X_2}$$

and finally use the molecular weights to establish the ratio by mass:

$$\frac{n_2}{n_1} = \frac{X_2}{1 - X_2} = \frac{0.459 \text{ mol } CH_3CH_2OH}{(1 - 0.459) \text{ mol } H_2O} \times \frac{46.069 \text{ g } CH_3CH_2OH}{\text{mol } CH_3CH_2OH} \times \frac{1 \text{ mol } H_2O}{18.015 \text{ g } H_2O}$$

$$= \frac{2.17 \text{ g } CH_3CH_2OH}{\text{g } H_2O}$$

Chapter 10

Macroscopic to Microscopic—Gases and Kinetic Theory

S10-1. Suppose that water is used to construct a barometer. Will a column of water that registers 1.0000 atm on a hot summer day be taller than, shorter than, or the same height as a column of water that registers 1.0000 atm on a freezing winter day? Consult the appropriate table in Appendix C for any necessary data.

S10-2. Suppose, instead, that ethanol is used to construct a barometer. Calculate the density of the liquid, given that a column of ethanol rises 42.9 ft when the ambient pressure is 1.00 atm at 20°C. Consult Appendix C for additional data, if necessary.

S10-3. Why do you think that mercury is a preferred material for barometers?

S10-4. Assume that an ideal gas is confined at STP. How many particles, on the average, are contained in spheres with the following radii?

(a) 1.00 m **(b)** 1.00 cm **(c)** 1.00 mm **(d)** 1.00 μm **(e)** 10.0 nm

S10-5. Charles's law states that

$$\frac{V}{T} = \text{constant}$$

for fixed n and P. **(a)** What is the value of the constant in Charles's law for $n = 1.0000$ mol and $P = 1.0000$ atm? **(b)** What is its value for $n = 0.10000$ mol and $P = 10.000$ atm? **(c)** For $n = 10.000$ mol and $P = 1.0000$ atm? **(d)** Is the value constant for all choices of n and P? Why or why not?

S10-6. Boyle's law states that

$$PV = \text{constant}$$

for fixed n and T. **(a)** Under what conditions will the constant in Boyle's law have the magnitude R (the universal gas constant)? **(b)** Under what conditions will it have the value 1.0000 atm L? **(c)** Under what conditions will it have the value 1.0000 J?

S10-7. Calculate the ratio of the constants k_{Boyle} and k_{Charles} in Boyle's law and Charles's law:

$$PV = k_{\text{Boyle}} \qquad (\text{fixed } n, T)$$

$$\frac{V}{T} = k_{\text{Charles}} \qquad (\text{fixed } n, P)$$

S10-8. The temperature of an ideal gas is increased from 200 K to 400 K. At the same time, the pressure is increased from 1.50 atm to 3.00 atm. What happens to the density—does it increase, decrease, or remain the same?

S10-9. The temperature of an ideal gas is decreased from 300°C to 150°C. At the same time, the pressure is decreased from 2.00 atm to 1.00 atm. What happens to the density—does it increase, decrease, or remain the same?

S10-10. A 50.0-g mass of gaseous O_2 fills a container at STP. **(a)** Calculate the volume of the container. **(b)** How many particles of Xe would be found in a container twice the volume?

S10-11. How many grams of gaseous CO_2 occupy a volume of 3.50 L at a pressure of 1.10 atm and a temperature of 25.0°C?

S10-12. A mixture of two ideal gases has a total pressure of 535.1 torr. If the mole fraction of one component is 0.327, what is the partial pressure of the other component?

S10-13. What additional information will you need to solve the following problem?

> A mixture of two ideal gases has a total pressure of 535.1 torr. If the mole fraction of one component is 0.327, what is the density of the other component?

S10-14. What additional information will you need to solve the following problem?

> A sample of ideal gas occupies a volume of 22.4 L at 1.00 atm. How many particles are present in the container?

S10-15. What additional information will you need to solve the following problem?

> A certain ideal gas consists of diatomic molecules X_2. Identify X, given that the total mass of a sample at STP is 10.00 g.

S10-16. What additional information will you need to solve the following problem?

> A 3.00-mol sample of an ideal gas has a pressure of 0.333 atm. Calculate the average translational kinetic energy per particle.

S10-17. The root-mean-square speed of He is 1531 m s^{-1} at a certain temperature. Calculate the temperature.

S10-18. The translational kinetic energy of H_2 is 4.00 kJ mol^{-1} at a certain temperature. Calculate the root-mean-square speed of He at the same temperature.

S10-19. A mixture of H_2, He, and N_2 occupies a volume of 11.2 L at STP. Calculate the average translational kinetic energy *per particle* that results for **(a)** H_2, **(b)** He, **(c)** N_2.

S10-20. A mixture of H_2, He, and N_2 occupies a volume of 44.8 L at STP. Calculate the root-mean-square speed that results for **(a)** H_2, **(b)** He, **(c)** N_2.

S10-21. A 20.0-L sample of neon gas, originally at STP, is compressed isothermally to a volume of 18.0 L. **(a)** Calculate the translational kinetic energy per mole after the compression. **(b)** Calculate the root-mean-square speed after the compression.

S10-22. The root-mean-square speed of a certain ideal gas falls by 50% in response to a halving of the volume. What is the corresponding change in pressure?

S10-23. Recall the definition of the *mean free path* for a gas particle: the average distance traveled between collisions. **(a)** Will the mean free path increase, decrease, or remain the same the same if P is doubled at constant T? **(b)** Will the mean free path increase, decrease, or remain the same if T is doubled at constant P? **(c)** Will the mean free path increase, decrease, or remain the same if n is halved at constant P and T?

S10-24. Consider, in anticipation of Chapter 11, an obvious flaw in our model of the ideal gas: The particles in a gas clearly *do* have intrinsic volume; atoms and molecules are not simply point masses. Assume, therefore, that each particle occupies a finite space but is otherwise unable to interact with any other particle. **(a)** Will the volume of this special kind of "real" gas be greater than, less than, or the same as the volume of an ideal gas? **(b)** Given that

$$P_{\text{ideal}}V_{\text{ideal}} = nRT$$

for an ideal gas, how would you expect the intrinsic particle volume to affect the pressure of a real gas (P_{real})? Will P_{real} be higher than, lower than, or the same as P_{ideal}? **(c)** Under what conditions of temperature and pressure will it be most reasonable to neglect the intrinsic volume of the gas particles?

S10-25. Consider another obvious flaw in our picture of an ideal gas: Atoms and molecules certainly do have the ability to interact. All the usual means of intermolecular interactions—hydrogen bonding, dipole–dipole, dispersion, and so forth—remain inherent in the internal structure of the particles. **(a)** How would you expect these intermolecular interactions to affect the pressure of a real gas? Will P_{real} be higher than, lower than, or the same as P_{ideal}? **(b)** Under what conditions of volume and temperature will it be most reasonable to neglect the interactions?

S10-26. For which gas in each pair do you expect the equation of state

$$PV = nRT$$

to be more accurate?

(a) He, Rn **(b)** H_2O, CH_4 **(c)** H_2, CO_2

S10-27. Take into account the preceding three exercises, and pick the statement that best describes the relationship between P_{real} and P_{ideal}:

(a) P_{real} is always greater than P_{ideal}.
(b) P_{real} is always less than P_{ideal}.
(c) P_{real} is always equal to P_{ideal}.
(d) P_{real} may be greater than, less than, or equal to P_{ideal}.

Justify your answer.

S10-28. Explain why Dalton's law of partial pressures is valid for ideal gases.

S10-29. The ideal gas equation of state predicts that the volume of a gas shrinks to zero as the temperature approaches 0 K:

$$V = \frac{nRT}{P}$$

(a) Why is this prediction *always* wrong for a real gas? What happens instead? **(b)** For which of these two gases—carbon dioxide or helium—will the "ideal" prediction be more nearly correct at a given temperature? Why?

SOLUTIONS

S10-1. Exercise 8 in Chapter 10 of *Principles of Chemistry* introduces the equation connecting pressure (P) with the density (ρ) and height (h) of a barometric column:

$$P = \rho g h \qquad (g = 9.81 \text{ m s}^{-2})$$

This relationship tells us, in particular, that the height attained for a given pressure is inversely proportional to the density:

$$h = \frac{P}{\rho g}$$

The higher the density, the lower the column.

Since water is slightly more dense at the lower temperature, a column that registers 1.0000 atm will be correspondingly shorter during the winter. See Table C-15 in the text.

S10-2. Substitute $P = 1.00$ atm and $h = 42.9$ ft into the equation $P = \rho g h$, and solve for the density ρ. The necessary unit conversions are worked out explicitly below:

$$\rho = \frac{P}{gh} = \frac{1.00 \text{ atm} \times \left(\dfrac{1.01325 \times 10^5 \text{ Pa}}{\text{atm}} \times \dfrac{1 \text{ N m}^{-2}}{\text{Pa}} \times \dfrac{1 \text{ kg m s}^{-2}}{\text{N}} \times \dfrac{1 \text{ m}}{100 \text{ cm}} \times \dfrac{1000 \text{ g}}{\text{kg}} \right)}{\left(981 \text{ cm s}^{-2}\right)\left(42.9 \text{ ft} \times \dfrac{12 \text{ in}}{\text{ft}} \times \dfrac{2.54 \text{ cm}}{\text{in}}\right)}$$

$$= 0.790 \text{ g cm}^{-3}$$

S10-3. Mercury is preferred for its high density, 13.595 g cm^{-3} at 0°C. A column of mercury rises to a height of only 760 mm when subjected to a pressure of 1 atm.

S10-4. Start with $PV = nRT$. Insert the volume of a sphere with radius r,

$$V = \tfrac{4}{3}\pi r^3$$

and then use Avogadro's number (N_0) to solve for N, the number of particles:

$$n = \frac{N}{N_0} = \frac{PV}{RT}$$

$$N = \frac{N_0 PV}{RT} = \frac{4\pi N_0 P r^3}{3RT}$$

All that remains is to convert m^3 to L, as demonstrated below for $r = 1.00$ m:

$$N = \frac{4\pi N_0 P r^3}{3RT} = \frac{4\pi\left(6.022\times10^{23}\ \text{mol}^{-1}\right)\left(1.00\ \text{atm}\right)\left[\left(1.00\ \text{m}\times\dfrac{100\ \text{cm}}{\text{m}}\right)^3\times\dfrac{1\ \text{L}}{1000\ \text{cm}^3}\right]}{3\left(0.08206\ \text{atm L mol}^{-1}\ \text{K}^{-1}\right)\left(273.15\ \text{K}\right)}$$

$$= 1.13\times10^{26}$$

Results for the various volumes are tabulated below:

	r	N	CONVERSION
(a)	1.00 m	1.13×10^{26}	$1\ \text{L} = 10^{-3}\ \text{m}^3$
(b)	1.00 cm	1.13×10^{20}	$1\ \text{L} = 10^{3}\ \text{cm}^3$
(c)	1.00 mm	1.13×10^{17}	$1\ \text{L} = 10^{6}\ \text{mm}^3$
(d)	1.00 μm	1.13×10^{8}	$1\ \text{L} = 10^{15}\ \mu\text{m}^3$
(e)	10.0 nm	1.13×10^{2}	$1\ \text{L} = 10^{24}\ \text{nm}^3$

S10-5. We know, from the ideal gas equation, that the constant in Charles's law is equal to nR/P at fixed n and P:

$$k_{\text{Charles}} = \frac{V}{T} = \frac{nR}{P} \qquad (\text{fixed } n,\ P)$$

(a) Insert the specified values of n and P to calculate k_{Charles}:

$$k_{\text{Charles}} = \frac{nR}{P} = \frac{\left(1.0000\ \text{mol}\right)\left(0.082058\ \text{atm L mol}^{-1}\ \text{K}^{-1}\right)}{1.0000\ \text{atm}} = 0.082058\ \text{L K}^{-1}$$

(b) Given different values for n and P, we obtain a different value for k_{Charles}:

$$k_{\text{Charles}} = \frac{nR}{P} = \frac{\left(0.10000\ \text{mol}\right)\left(0.082058\ \text{atm L mol}^{-1}\ \text{K}^{-1}\right)}{10.000\ \text{atm}} = 8.2058\times10^{-4}\ \text{L K}^{-1}$$

(c) Similar:

$$k_{\text{Charles}} = \frac{nR}{P} = \frac{\left(10.000\ \text{mol}\right)\left(0.082058\ \text{atm L mol}^{-1}\ \text{K}^{-1}\right)}{1.0000\ \text{atm}} = 0.82058\ \text{L K}^{-1}$$

(d) The "constant" k_{Charles} depends on the ratio n/P. It has the same value only for those combinations of n and P that yield the same quotient.

See the next exercise for a related treatment of Boyle's law.

S10-6. The approach is similar to that taken in Exercise S10-5. Using the ideal gas equation, we see at once that the constant in Boyle's law is equal to nRT at fixed n and T:

$$k_{\text{Boyle}} = PV = nRT \qquad (\text{fixed } n, T)$$

(a) The magnitude of k_{Boyle} becomes equal to R whenever nT is equal to 1.0000 mol K. Acceptable units of k_{Boyle} are atm L, representing pressure × volume (work or energy):

$$k_{\text{Boyle}} = nRT = R \text{ atm L} = 0.082058 \text{ atm L} \quad (\text{given})$$

$$nT = \frac{k_{\text{Boyle}}}{R} = \frac{0.082058 \text{ atm L}}{0.082058 \text{ atm L mol}^{-1}\text{ K}^{-1}} = 1.0000 \text{ mol K}$$

(b) Use the value

$$R = 0.082058 \text{ atm L mol}^{-1}\text{ K}^{-1}$$

to show that $k_{\text{Boyle}} = 1.0000$ atm L whenever $nT = 1/R$:

$$k_{\text{Boyle}} = nRT = 1.0000 \text{ atm L} \quad (\text{given})$$

$$nT = \frac{k_{\text{Boyle}}}{R} = \frac{1.0000 \text{ atm L}}{0.082058 \text{ atm L mol}^{-1}\text{ K}^{-1}} = 12.187 \text{ mol K}$$

(c) Similar, but here we insert the value $R = 8.3145 \text{ J mol}^{-1}\text{ K}^{-1}$:

$$k_{\text{Boyle}} = nRT = 1.0000 \text{ J} \quad (\text{given})$$

$$nT = \frac{k_{\text{Boyle}}}{R} = \frac{1.0000 \text{ J}}{8.3145 \text{ J mol}^{-1}\text{ K}^{-1}} = 0.12027 \text{ mol K}$$

S10-7. Combine the results of Exercises S10-5 and S10-6:

$$\frac{k_{\text{Boyle}}}{k_{\text{Charles}}} = \frac{nRT}{\frac{nR}{P}} = PT$$

S10-8. If both the pressure and absolute temperature are doubled,

$$P_1 = 1.50 \text{ atm} \qquad T_1 = 200 \text{ K}$$

$$P_2 = 3.00 \text{ atm} \qquad T_2 = 400 \text{ K}$$

the density (ρ) remains the same:

$$PV = nRT$$

$$\rho_2 = \frac{n_2}{V_2} = \frac{P_2}{RT_2} = \frac{2P_1}{R(2T_1)} = \frac{P_1}{RT_1} = \rho_1$$

S10-9. The pressure is cut in half, but a change from 300°C to 150°C does not correspond to a halving of the *Kelvin* temperature:

$$P_1 = 2.00 \text{ atm} \qquad T_1 = (300 + 273.15) \text{ K} = 573 \text{ K}$$

$$P_2 = 1.00 \text{ atm} \qquad T_2 = (150 + 273.15) \text{ K} = 423 \text{ K}$$

The density therefore decreases by approximately 32%:

$$\frac{P_1V_1}{n_1T_1} = \frac{P_2V_2}{n_2T_2}$$

$$\frac{n_2}{V_2} = \frac{n_1}{V_1} \times \frac{P_2}{P_1} \times \frac{T_1}{T_2}$$

$$\rho_2 = \rho_1 \times \frac{P_2}{P_1} \times \frac{T_1}{T_2} = \rho_1 \times \frac{1.00 \text{ atm}}{2.00 \text{ atm}} \times \frac{573 \text{ K}}{423 \text{ K}} = 0.677\rho_1$$

S10-10. We are given values for the pressure, temperature, and total mass of a sample of gaseous oxygen:

$$P = 1.00 \text{ atm} \qquad T = 273 \text{ K} \qquad m_{\text{tot}} = 50.0 \text{ g}$$

(a) Using the molar mass to convert grams into moles, we have sufficient information to calculate the volume:

$$V = \frac{nRT}{P} = \frac{\left(50.0 \text{ g O}_2 \times \frac{1 \text{ mol}}{31.9988 \text{ g O}_2}\right)(0.08206 \text{ atm L mol}^{-1} \text{ K}^{-1})(273 \text{ K})}{1.00 \text{ atm}} = 35.0 \text{ L}$$

(b) Avogadro's law, applicable to systems at the same temperature and pressure, states that equal volumes of ideal gas contain equal numbers of particles. If the volume is doubled, then the number of particles—whether Xe or O_2—will also be doubled. Chemical identity is irrelevant, provided that the system behaves ideally:

$$\left(50.0 \text{ g O}_2 \times \frac{1 \text{ mol O}_2}{31.9988 \text{ g O}_2}\right) \times \frac{1 \text{ mol Xe}}{\text{mol O}_2} \times \frac{6.022 \times 10^{23} \text{ atoms}}{\text{mol}} \times \frac{70.0 \text{ L}}{35.0 \text{ L}}$$

$$= 1.88 \times 10^{24} \text{ atoms Xe}$$

S10-11. Express the molar amount in terms of the total mass (m_{tot}) and molar mass ($\mathcal{M}$),

$$n = \frac{m_{tot}}{\mathcal{M}}$$

and rearrange $PV = nRT$ to solve for m_{tot}:

$$m_{tot} = \frac{\mathcal{M}PV}{RT} = \frac{(44.010 \text{ g CO}_2 \text{ mol}^{-1})(1.10 \text{ atm})(3.50 \text{ L})}{(0.08206 \text{ atm L mol}^{-1} \text{ K}^{-1})(298.15 \text{ K})} = 6.93 \text{ g CO}_2$$

S10-12. The sum of the mole fractions, X_1 and X_2, is equal to 1. Given the mole fraction of one component and the total pressure of the mixture,

$$X_1 = 0.327 \qquad P_{tot} = 535.1 \text{ torr}$$

we apply Dalton's law to compute the partial pressure of the other component:

$$P_2 = X_2P_{tot} = (1 - X_1)P_{tot} = (1 - 0.327)(535.1 \text{ torr}) = 360. \text{ torr}$$

S10-13. We have already shown, just above, that the information given is sufficient to establish the partial pressure of each component:

$$X_1 = 0.327 \qquad X_2 = 1 - X_1 \qquad P_{tot} = 535.1 \text{ torr}$$

$$P_i = X_iP_{tot} \qquad (i = 1, 2)$$

To determine the density, however, we need to know the temperature as well:

$$\rho_i = \frac{n_i}{V} = \frac{P_i}{RT}$$

S10-14. To calculate the number of moles (and hence the number of particles), we need to know the temperature in addition to the pressure and volume:

$$P = 1.00 \text{ atm} \qquad V = 22.4 \text{ L} \qquad T = \text{missing}$$

$$n \overset{?}{=} \frac{PV}{RT}$$

S10-15. To determine the identifying molar mass ($\mathcal{M}$), we need to know the volume in which a total mass (m_{tot}) is confined at a given temperature and pressure:

$$P = 1.00 \text{ atm} \qquad V = \text{missing} \qquad T = 273 \text{ K} \qquad m_{tot} = 10.00 \text{ g}$$

$$n = \frac{m_{tot}}{\mathcal{M}} = \frac{PV}{RT}$$

$$\mathcal{M} \overset{?}{=} \frac{m_{tot}RT}{PV}$$

S10-16. We need to know the temperature in order to determine the average kinetic energy:

$$\langle \varepsilon_k \rangle = \frac{3}{2} k_B T$$

Given n and P, however, we can compute T if we have the volume as well:

$$T = \frac{PV}{nR}$$

Thus either temperature or volume will suffice to establish the average kinetic energy per particle in this system.

S10-17. Manipulation of the equation

$$v_{rms} = \sqrt{\frac{3RT}{\mathcal{M}}}$$

gives us the temperature as a function of the root-mean square speed:

$$(v_{rms})^2 = \frac{3RT}{\mathcal{M}}$$

$$T = \frac{\mathcal{M}(v_{rms})^2}{3R} = \frac{\left(\frac{4.0026 \text{ g}}{\text{mol}} \times \frac{1 \text{ kg}}{1000 \text{ g}}\right)(1531 \text{ m s}^{-1})^2}{3(8.3145 \text{ kg m}^2 \text{ s}^{-2} \text{ mol}^{-1} \text{ K}^{-1})} = 376.1 \text{ K}$$

Recall that $1 \text{ J} = 1 \text{ kg m}^2 \text{ s}^{-2}$.

S10-18. First, rearrange the expression

$$E_{\rm k} = \frac{3}{2}RT$$

to obtain the temperature T in terms of the molar kinetic energy $E_{\rm k}$:

$$T = \frac{2E_{\rm k}}{3R}$$

Second, insert this form into the equation for the root-mean-square speed $v_{\rm rms}$:

$$v_{\rm rms} = \sqrt{\frac{3RT}{\mathcal{m}}} = \sqrt{\frac{2E_{\rm k}}{\mathcal{m}}} = \sqrt{\frac{2\left(4.00 \times 10^{3}\ \text{kg m}^2\ \text{s}^{-2}\ \text{mol}^{-1}\right)}{4.003 \times 10^{-3}\ \text{kg mol}^{-1}}} = 1.41 \times 10^{3}\ \text{m s}^{-1}$$

The conversions

$$1\ \text{J} = 1\ \text{N m} = 1\ \text{kg m}^2\ \text{s}^{-2} \qquad 1\ \text{kJ} = 1000\ \text{J} \qquad 1\ \text{kg} = 1000\ \text{g}$$

are implicit in the calculation.

S10-19. Let N denote the total number of particles, and let n denote the corresponding number of moles. Avogadro's number, N_0, relates one quantity to the other:

$$N = nN_0$$

With that, we have the average translational energy per particle—the total energy $E_{\rm k}$ divided by the number of particles N:

$$\langle \varepsilon_{\rm k} \rangle = \frac{E_{\rm k}}{N} = \frac{E_{\rm k}}{nN_0} = \frac{\frac{3}{2}nRT}{nN_0} = \frac{3}{2}k_{\rm B}T$$

By convention, the symbol $k_{\rm B}$ represents Boltzmann's constant:

$$k_{\rm B} = \frac{R}{N_0} = 1.38066 \times 10^{-23}\ \text{J K}^{-1}$$

The average translational energy per particle, dependent only on the temperature, is therefore the same for each of the component gases:

$$\langle \varepsilon_{\rm k} \rangle = \frac{3}{2}\left(1.38066 \times 10^{-23}\ \text{J K}^{-1}\right)\left(273\ \text{K}\right) = 5.65 \times 10^{-21}\ \text{J} \quad \text{(per particle)}$$

Note, moreover, that the specification of pressure and volume alone is insufficient to fix a value for $\langle \varepsilon_k \rangle$:

$$E_k = \frac{3}{2} nRT = \frac{3}{2} PV$$

$$N = nN_0 = \frac{PV}{RT} N_0$$

$$\frac{E_k}{N} = \frac{\left(\frac{3}{2} PV\right)}{\left(\frac{PVN_0}{RT}\right)} = \frac{3}{2}\left(\frac{R}{N_0}\right)T = \frac{3}{2} k_B T$$

We need to supplement P and V with either n or T in order to determine $\langle \varepsilon_k \rangle$.

		$\langle \varepsilon_k \rangle$ (J)
(a)	H_2	5.65×10^{-21}
(b)	He	5.65×10^{-21}
(c)	N_2	5.65×10^{-21}

S10-20. The molar translational energy

$$E_k = \frac{3}{2} RT$$

is the same for each component, but particles of different masses develop different root-mean-square speeds. The sample calculation below is for H_2, with a molar mass of 2.0159 g mol^{-1}:

$$v_{rms} = \sqrt{\frac{3RT}{\mathcal{M}}} = \sqrt{\frac{3(8.3145 \text{ J mol}^{-1} \text{ K}^{-1})(273 \text{ K})}{2.0159 \times 10^{-3} \text{ kg mol}^{-1}}} = 1.84 \times 10^3 \text{ m s}^{-1}$$

$$(1 \text{ J} = 1 \text{ kg m}^2 \text{ s}^{-2})$$

The more massive particles in the mixture move at slower average speeds:

		$\mathcal{M}$ (g mol^{-1})	v_{rms} (m s^{-1})
(a)	H_2	2.0159	1.84×10^3
(b)	He	4.0026	1.30×10^3
(c)	N_2	28.013	4.93×10^2

S10-21. The temperature is unchanged by any isothermal compression: $T_2 = T_1 = 273$ K.

(a) Knowing the temperature, we also know the molar kinetic energy:

$$E_k = \frac{3}{2}RT = \frac{3}{2}(8.3145 \text{ J mol}^{-1} \text{ K}^{-1})(273 \text{ K}) = 3.40 \times 10^3 \text{ J mol}^{-1}$$

(b) The root-mean-square speed, too, is fixed by the temperature:

$$v_{\text{rms}} = \sqrt{\frac{3RT}{\mathcal{M}}} = \sqrt{\frac{3(8.3145 \text{ J mol}^{-1} \text{ K}^{-1})(273 \text{ K})}{20.1797 \times 10^{-3} \text{ kg mol}^{-1}}} = 581 \text{ m s}^{-1}$$

Once again, we make implicit use of the relationship 1 J = 1 N m = 1 kg $\text{m}^2 \text{ s}^{-2}$. The molar mass is likewise expressed in kilograms, consistent with the units of R.

S10-22. Reducing the *temperature* by a factor of 4 causes the root-mean-square speed to decrease by a factor of 2:

$$\frac{v_{\text{rms}}(T_2)}{v_{\text{rms}}(T_1)} = \sqrt{\frac{T_2}{T_1}} = \frac{1}{2} \quad \text{(given)}$$

$$\frac{T_2}{T_1} = \left[\frac{v_{\text{rms}}(T_2)}{v_{\text{rms}}(T_1)}\right]^2 = \left(\frac{1}{2}\right)^2 = \frac{1}{4}$$

We are told, furthermore, that the volume is cut in half,

$$\frac{V_2}{V_1} = \frac{1}{2}$$

and thus the ideal gas law requires that the pressure be cut in half as well:

$$\frac{P_1V_1}{T_1} = \frac{P_2V_2}{T_2}$$

$$P_2 = P_1 \times \frac{V_1}{V_2} \times \frac{T_2}{T_1} = P_1 \times 2 \times \frac{1}{4} = \frac{P_1}{2}$$

S10-23. The higher the density,

$$\rho = \frac{n}{V} = \frac{P}{RT}$$

the smaller is the mean free path.

(a) The density doubles when the pressure is doubled at constant temperature:

$$P_2 = 2P_1 \qquad\qquad T_2 = T_1$$

$$\frac{\rho_2}{\rho_1} = \frac{P_2 T_1}{P_1 T_2} = \frac{(2P_1)T_1}{P_1 T_1} = 2$$

With the particles crowded into a smaller volume, the mean free path decreases correspondingly.

(b) A doubling of the temperature at constant pressure causes the opposite effect: a *halving* of the density.

$$P_2 = P_1 \qquad\qquad T_2 = 2T_1$$

$$\frac{\rho_2}{\rho_1} = \frac{P_2 T_1}{P_1 T_2} = \frac{P_1 T_1}{P_1(2T_1)} = \frac{1}{2}$$

The mean free path goes up.

(c) The density—and hence the mean free path—remains unchanged whenever the ratio P/T holds constant. A halving of n at constant P and T is accompanied by a halving of V, leaving ρ the same before and after:

$$\frac{\rho_2}{\rho_1} = \frac{\dfrac{n_2}{V_2}}{\dfrac{n_1}{V_1}} = \frac{\dfrac{\frac{1}{2}n_1}{\frac{1}{2}V_1}}{\dfrac{n_1}{V_1}} = 1$$

S10-24. Here, and in the remaining supplementary exercises for this chapter, we consider more closely what it means for a gas to be ideal. The subject of real gases is taken up immediately following in Chapter 11.

(a) If indeed each particle occupies a finite volume, then the amount of free space open to the particles will be less than the volume of the empty container. The effective volume of this kind of *real* gas is smaller than the volume of an ideal gas.

(b) We assume that pressure and volume are inversely proportional:

$$P \propto \frac{1}{V}$$

If $V_{\text{real}} < V_{\text{ideal}}$, then P_{real} will be greater than P_{ideal}. The larger the intrinsic particle volume, the greater will be the increase of P_{real} relative to P_{ideal}.

(c) The intrinsic particle volume becomes a smaller fraction of the total volume when the density is low, as happens at high temperature and low pressure.

S10-25. A preview of the relationship between intermolecular interactions and pressure in a real gas.

(a) To the extent that particles cluster together inside the container (away from the walls), the frequency of collisions with the walls will decrease. The real pressure is therefore expected to be less than the nominal ideal pressure.

(b) Interactions become less important at low densities: high volume, high temperature. First, collisions grow less likely in a large, sparsely populated space. Second, when collisions do occur, particles with high kinetic energy are better able to overcome the relatively weak forces of attraction.

S10-26. The weaker the interparticle interactions, the more nearly ideal the gas will be.

(a) He, less massive than Rn, experiences a weaker London interaction. Helium consequently behaves more like an ideal gas.

(b) H_2O interacts by means of comparatively strong hydrogen bonds, whereas CH_4 interacts primarily by means of the weak London interaction. Methane gas better approaches the limit of ideality.

(c) London interactions are weaker for H_2, a molecule less massive than CO_2. Hydrogen gas is more nearly ideal.

S10-27. Choice (d) is correct: P_{real} may be greater than, less than, or equal to P_{ideal}. The outcome depends on the interplay of competing factors, such as the effect of excluded volume (which tends to increase the real pressure) and the effect of intermolecular interactions (which tends to decrease the real pressure).

See Exercises S10-24 and S10-25, as well as Section 11-2, Examples 11-1 and 11-2, and the accompanying exercises in *Principles of Chemistry*.

S10-28. Dalton's law presupposes that different kinds of particles interact in exactly the same way. This condition is fulfilled in an ideal gas, where indeed the interactions are all the same—zero—regardless of the kind of particle.

S10-29. No gas is truly ideal, and no gas can remain a gas at sufficiently low temperature and high pressure.

(a) The particles move slower as the temperature decreases, and hence the average kinetic energy decreases as well. At some point, the thermal energy becomes low enough for the particles to interact effectively and form clusters. The gas condenses into a liquid.

(b) Helium, which interacts more weakly than carbon dioxide, remains an "ideal" gas over a wider range of temperature. Nevertheless, He does condense under the appropriate conditions.

Chapter 11

Disorder–Order and Phase Transitions

S11-1. Is it possible for a *real* gas to remain a gas as the temperature approaches absolute zero, regardless of density?

S11-2. Suppose that 25.0 grams of gaseous methane (CH_4) fill a volume of 100.0 L at a pressure of 1.00 atm. Is it possible to force a condensation by reducing the volume below 100 L? Note that the critical temperature for methane is –82.62°C.

S11-3. The density and vapor pressure of diethyl ether ($C_4H_{10}O$) at 25°C are 0.7080 g mL^{-1} and 532.5 torr, respectively. Calculate the vapor pressure of a solution in which 0.500 gram of naphthalene ($C_{10}H_8$) is dissolved in 25.0 milliliters of diethyl ether.

S11-4. Calculate the total vapor pressure over a solution that contains equal volumes of benzene (C_6H_6) and toluene ($C_6H_5CH_3$) at 25°C, given the information below:

COMPONENT	DENSITY (g mL^{-1})	VAPOR PRESSURE (torr, 25°C)
benzene	0.8729	95.3
toluene	0.8647	28.5

S11-5. Calculate the total vapor pressure over a solution that contains equal masses of methanol (CH_3OH) and water at 25°C, given the information below:

COMPONENT	DENSITY (g mL^{-1})	VAPOR PRESSURE (torr, 25°C)
methanol	0.7872	127.5
water	0.99705	23.8

S11-6. The vapor pressure of X (a solvent) is lowered by 10.0% when 1.00 mole of Y (a molecular solute) is dissolved in 1.00 kilogram of X. Do you have sufficient data to determine the molar mass of Y? If not, what additional information is needed?

S11-7. An aqueous solution of NaBr freezes at a temperature of −1.15°C. Calculate the mass of dissolved NaBr per kilogram of solvent.

S11-8. An aqueous solution contains 5.00 g $Ca(NO_3)_2$ and boils at 374.15°C. Calculate the corresponding mass of solvent.

S11-9. An aqueous solution contains 15.00 g $Ce_2(SO_4)_3$ dissolved in 150. mL H_2O. Assume that the density of water is 1.00 g mL^{-1}. **(a)** Calculate the freezing point. **(b)** How many grams of $NaCH_3COO$ must be dissolved in 400. mL H_2O to produce the same freezing point?

S11-10. A particular aqueous solution of sucrose freezes at −0.75°C. At what temperature does it boil?

S11-11. Suppose that a certain aqueous solution freezes at −3.00°C. **(a)** Can you determine the corresponding boiling point without knowing whether the solute is molecular or ionic? **(b)** If so, do it: Calculate the boiling point.

S11-12. The density of acetone (C_3H_6O) is 0.7899 g mL^{-1} at 20°C. When 2.00 grams of naphthalene ($C_{10}H_8$) are dissolved in 20.0 milliliters of acetone, the boiling point rises from 56.0°C to 57.7°C. Calculate the molal boiling-point-elevation constant, K_b, for acetone.

S11-13. The freezing point of water falls by 0.412°C when 1.00 g of an unknown molecular solute is dissolved. Given the appropriate value of K_f, do you have sufficient data to determine the molar mass of the solute? If not, what additional information is needed?

S11-14. The boiling point of methanol rises by 0.628°C when 0.562 mol of a molecular solute is dissolved. Given the appropriate value of K_b, do you have sufficient data to determine the density of methanol? If not, what additional information is required?

S11-15. Make a rough sketch, from memory, of the phase diagram for a substance such as CO_2. Mark the following locations:

(a) solid, liquid, and gas regions
(b) solid–liquid, solid–gas, and gas–liquid coexistence regions
(c) critical isotherm
(d) supercritical region

S11-16. Locate the following points on the phase diagram requested in Exercise S11-15:

(a) triple point
(b) critical point
(c) normal melting point
(d) normal boiling point

S11-17. Sketch, again from memory, the following transitions on the phase diagram used in Exercises S11-15 and S11-16:

(a) freezing
(b) melting
(c) condensation
(d) vaporization
(e) deposition
(f) sublimation

Represent each transition as both an isothermal and an isobaric process.

S11-18. Sketch the phase diagram for a substance that is more dense in its solid form than in its liquid form.

S11-19. Sketch the phase diagram for a substance that undergoes an increase in molar volume upon freezing.

S11-20. The vapor pressure for substance 1 varies sharply with temperature, whereas for substance 2 the variation is far less pronounced. Sketch the phase diagram for each substance.

S11-21. Does the boiling temperature of water increase, decrease, or remain the same as the ambient pressure decreases? Show the effect on a phase diagram.

S11-22. Which is the ordered phase and which is the disordered phase in each pair?

(a) $CH_3CH_2OH(g)$, $CH_3CH_2OH(\ell)$
(b) $CH_3CH_2OH(s)$, $CH_3CH_2OH(\ell)$
(c) $NaCl(s)$, $NaCl(aq)$
(d) $H_2O(\ell)$ at 25°C, $H_2O(\ell)$ at 30°C

SOLUTIONS

S11-1. No. All real gases can be made to condense below a certain critical temperature T_c, under conditions where the thermal energy becomes comparable to the intermolecular potential.

S11-2. This system of methane gas will not condense under the conditions specified, no matter how much pressure is applied. The temperature, as determined by the ideal gas equation, is greater than T_c:

$$T = \frac{PV}{nR} = \frac{(1.00\ \text{atm})(100.0\ \text{L})}{\left(25.0\,\text{g CH}_4 \times \dfrac{1\ \text{mol}}{16.043\ \text{g CH}_4}\right)(0.08206\ \text{atm L mol}^{-1}\ \text{K}^{-1})} = 782\ \text{K}$$

$$T_c = (-82.62 + 273.15)\ \text{K} = 190.53\ \text{K}$$

S11-3. From the molar amounts of solvent and solute,

$$n_{C_4H_{10}O} = 25.0\ \text{mL C}_4\text{H}_{10}\text{O} \times \frac{0.7080\ \text{g C}_4\text{H}_{10}\text{O}}{\text{mL C}_4\text{H}_{10}\text{O}} \times \frac{1\ \text{mol}}{74.123\ \text{g C}_4\text{H}_{10}\text{O}} = 0.239\ \text{mol}$$

$$n_{C_{10}H_8} = 0.500\ \text{g C}_{10}\text{H}_8 \times \frac{1\ \text{mol}}{128.174\ \text{g C}_{10}\text{H}_8} = 0.00390\ \text{mol}$$

we calculate the mole fraction of solvent

$$X_{C_4H_{10}O} = \frac{n_{C_4H_{10}O}}{n_{C_4H_{10}O} + n_{C_{10}H_8}}$$

and we apply Raoult's law:

$$P_{C_4H_{10}O} = X_{C_4H_{10}O} P^{\circ}_{C_4H_{10}O} = \left(\frac{0.239\ \text{mol}}{0.239\ \text{mol} + 0.00390\ \text{mol}}\right)(532.5\ \text{torr}) = 524\ \text{torr}$$

S11-4. Assume that the mixture contains 1.00 mL of each liquid:

$$n_{C_6H_6} = 1.00\ \text{mL C}_6\text{H}_6 \times \frac{0.8729\ \text{g C}_6\text{H}_6}{\text{mL C}_6\text{H}_6} \times \frac{1\ \text{mol}}{78.114\ \text{g C}_6\text{H}_6} = 0.01117\ \text{mol}$$

$$n_{C_6H_5CH_3} = 1.00\ \text{mL C}_6\text{H}_5\text{CH}_3 \times \frac{0.8647\ \text{g C}_6\text{H}_5\text{CH}_3}{\text{mL C}_6\text{H}_5\text{CH}_3} \times \frac{1\ \text{mol}}{92.141\ \text{g C}_6\text{H}_5\text{CH}_3} = 0.00938\ \text{mol}$$

The total vapor pressure is equal to the sum of the partial pressures, each given separately by Raoult's law:

$$P = X_{C_6H_6} P^{\circ}_{C_6H_6} + X_{C_6H_5CH_3} P^{\circ}_{C_6H_5CH_3}$$

$$= \left(\frac{0.01117 \text{ mol}}{0.01117 \text{ mol} + 0.00938 \text{ mol}}\right)(95.3 \text{ torr}) + \left(\frac{0.00938 \text{ mol}}{0.01117 \text{ mol} + 0.00938 \text{ mol}}\right)(28.5 \text{ torr})$$

$$= 64.8 \text{ torr}$$

S11-5. Similar, except here we take 1.00 g of each component liquid:

$$n_{CH_3OH} = 1.00 \text{ g } CH_3OH \times \frac{1 \text{ mol}}{32.042 \text{ g } CH_3OH} = 0.03121 \text{ mol}$$

$$n_{H_2O} = 1.00 \text{ g } H_2O \times \frac{1 \text{ mol}}{18.015 \text{ g } H_2O} = 0.05551 \text{ mol}$$

The combined vapor pressure, an average weighted by the mole fractions, is equal to 61.1 torr:

$$P = X_{CH_3OH} P^{\circ}_{CH_3OH} + X_{H_2O} P^{\circ}_{H_2O}$$

$$= \left(\frac{0.03121 \text{ mol}}{0.03121 \text{ mol} + 0.05551 \text{ mol}}\right)(127.5 \text{ torr}) + \left(\frac{0.05551 \text{ mol}}{0.03121 \text{ mol} + 0.05551 \text{ mol}}\right)(23.8 \text{ torr})$$

$$= 61.1 \text{ torr}$$

S11-6. The information given is insufficient to calculate $\mathcal{m}_Y$, the molar mass of Y. We can determine $\mathcal{m}_X$ (see below),

$$P_X = X_X P^{\circ}_X = \left(\frac{n_X}{n_X + n_Y}\right) P^{\circ}_X = \left(\frac{n_X}{n_X + 1.00 \text{ mol}}\right) P^{\circ}_X = 0.900 P^{\circ}_X$$

$$n_X = 9.00 \text{ mol}$$

$$\mathcal{m}_X = \frac{1.00 \text{ kg X} \times \dfrac{1000 \text{ g}}{\text{kg}}}{9.00 \text{ mol}} = 111 \text{ g X mol}^{-1}$$

but to determine $\mathcal{m}_Y$ we need to know the mass of Y corresponding to $n_Y = 1.00$ mol.

S11-7. From the freezing-point depression,

$$\Delta T_f = K_f m = 1.15°\text{C}$$

we determine first the molality of solution, *m* (not to be confused with *mass*):

$$m = \frac{\Delta T_f}{K_f} = \frac{1.15°\text{C}}{1.86°\text{C kg mol}^{-1}} = \frac{0.618 \text{ mol dissolved ions}}{\text{kg H}_2\text{O}}$$

From that, we have the corresponding mass of solute:

$$\frac{0.618 \text{ mol dissolved ions}}{\text{kg H}_2\text{O}} \times \frac{1 \text{ mol NaBr}}{2 \text{ mol dissolved ions}} \times \frac{102.894 \text{ g NaBr}}{\text{mol NaBr}} = \frac{31.8 \text{ g NaBr}}{\text{kg H}_2\text{O}}$$

S11-8. We need to divide the molar amount of dissolved ions (n_{solute}) by the molality of solution (m) to obtain the mass of solvent:

$$\frac{n_{\text{solute}}}{m} \sim \text{mol solute} \times \frac{\text{kg solvent}}{\text{mol solute}} = \text{kg solvent}$$

Calculating the total molality from the specified boiling-point elevation,

$$m = \frac{\Delta T_b}{K_b} = \frac{(374.15°\text{C} - 373.15°\text{C})}{0.51°\text{C kg / mol ions}} = \frac{1.96 \text{ mol dissolved ions}}{\text{kg solvent}}$$

we thus determine the mass of solvent (H_2O):

$$\left(5.00 \text{ g Ca(NO}_3)_2 \times \frac{1 \text{ mol Ca(NO}_3)_2}{164.088 \text{ g Ca(NO}_3)_2} \times \frac{3 \text{ mol ions}}{\text{mol Ca(NO}_3)_2}\right) \times \frac{1 \text{ kg H}_2\text{O}}{1.96 \text{ mol ions}} = 0.047 \text{ kg H}_2\text{O}$$

The two-digit accuracy of K_b limits the final result to two significant figures.

S11-9. More practice with freezing-point depression. Remember throughout that colligative effects depend *collectively* on the total concentration of dissolved particles.

(a) Cerium(III) sulfate, $Ce_2(SO_4)_3$, releases five moles of ions per mole of solute dissolved:

$$\Delta T_f = K_f m$$

$$= \left(1.86^\circ\text{C kg mol}^{-1}\right)\left(\frac{15.00 \text{ g Ce}_2(\text{SO}_4)_3 \times \dfrac{1 \text{ mol}}{568.423 \text{ g}} \times \dfrac{5 \text{ mol ions}}{\text{mol}}}{150. \text{ mL H}_2\text{O} \times \dfrac{1.00 \text{ g}}{\text{mL}} \times \dfrac{1 \text{ kg}}{1000 \text{ g}}}\right) = 1.636^\circ\text{C}$$

The freezing point falls to −1.64°C.

(b) Sodium acetate, $NaCH_3COO$, releases two moles of ions per mole of solute dissolved. Working backward from the required molality,

$$m = \frac{\Delta T_f}{K_f} = \frac{1.636^\circ\text{C}}{1.86^\circ\text{C kg mol}^{-1}} = 0.880 \text{ mol kg}^{-1}$$

we obtain the mass needed for the specified volume:

$$\frac{0.880 \text{ mol ions}}{1000 \text{ g H}_2\text{O}} \times \frac{1 \text{ mol NaCH}_3\text{COO}}{2 \text{ mol ions}} \times \frac{82.034 \text{ g NaCH}_3\text{COO}}{\text{mol NaCH}_3\text{COO}}$$

$$\times \frac{1.00 \text{ g H}_2\text{O}}{\text{mL H}_2\text{O}} \times 400. \text{ mL H}_2\text{O} = 14.4 \text{ g NaCH}_3\text{COO}$$

S11-10. Determine, first, the molality of solution from the freezing-point depression,

$$\Delta T_f = K_f m$$

$$m = \frac{\Delta T_f}{K_f}$$

and then calculate the corresponding boiling-point elevation:

$$\Delta T_b = K_b m = K_b\left(\frac{\Delta T_f}{K_f}\right) = \left(0.51^\circ\text{C kg mol}^{-1}\right)\left(\frac{0.75^\circ\text{C}}{1.86^\circ\text{C kg mol}^{-1}}\right) = 0.21^\circ\text{C}$$

The boiling point rises to 100.21°C.

S11-11. Given the freezing point of a solution, can we calculate its boiling point as well?

(a) Yes. The key quantity is the effective molality,

$$m = \frac{\Delta T_f}{K_f}$$

which reflects the total concentration of dissolved particles—ions or molecules. This same value carries forward into the calculation of boiling point:

$$\Delta T_b = K_b m = K_b\left(\frac{\Delta T_f}{K_f}\right)$$

(b) Plug in the numbers:

$$\Delta T_b = \left(0.51^\circ\text{C kg mol}^{-1}\right)\left(\frac{3.00^\circ\text{C}}{1.86^\circ\text{C kg mol}^{-1}}\right) = 0.82^\circ\text{C}$$

The solution boils at 100.82°C.

S11-12. Given sufficient information to calculate ΔT_b and m,

$$\Delta T_b = 57.7^\circ\text{C} - 56.0^\circ\text{C} = 1.7^\circ\text{C}$$

$$m = \frac{2.00\text{ g } C_{10}H_8 \times \dfrac{1\text{ mol } C_{10}H_8}{128.174\text{ g } C_{10}H_8}}{20.0\text{ mL } C_3H_6O \times \dfrac{0.7899\text{ g } C_3H_6O}{\text{mL } C_3H_6O} \times \dfrac{1\text{ kg}}{1000\text{ g}}} = \frac{0.988\text{ mol } C_{10}H_8}{\text{kg } C_3H_6O}$$

we readily determine K_b:

$$K_b = \frac{\Delta T_b}{m} = \frac{1.7^\circ\text{C}}{0.988\text{ mol kg}^{-1}} = 1.7^\circ\text{C kg mol}^{-1}$$

S11-13. The information given is insufficient to calculate $\mathcal{M}_{\text{solute}}$, the molar mass of the solute. We can determine the molality of solution from ΔT_f and K_f,

$$\text{Molality} = \frac{\text{moles of solute}}{\text{mass of solvent}} = \frac{\Delta T_f}{K_f}$$

but to determine the moles of solute (and, from that, $\mathcal{M}_{\text{solute}}$) we need to know the mass of solvent into which the 1.00 gram is dissolved.

S11-14. The information given is insufficient to fix a value for the density of methanol, the solvent. We can determine the molality of solution from the boiling-point elevation,

$$\text{Molality} = \frac{\Delta T_b}{K_b}$$

and we can also determine the mass of methanol from the stated molar amount of solute:

$$\text{Molality} = \frac{\text{moles of solute}}{\text{mass of solvent } (CH_3OH)}$$

$$\text{Mass of solvent } (CH_3OH) = \frac{\text{moles of solute}}{\text{molality}}$$

To establish the density, however, we need to know the volume of methanol into which the solute is dissolved.

S11-15. See the schematic phase diagram below:

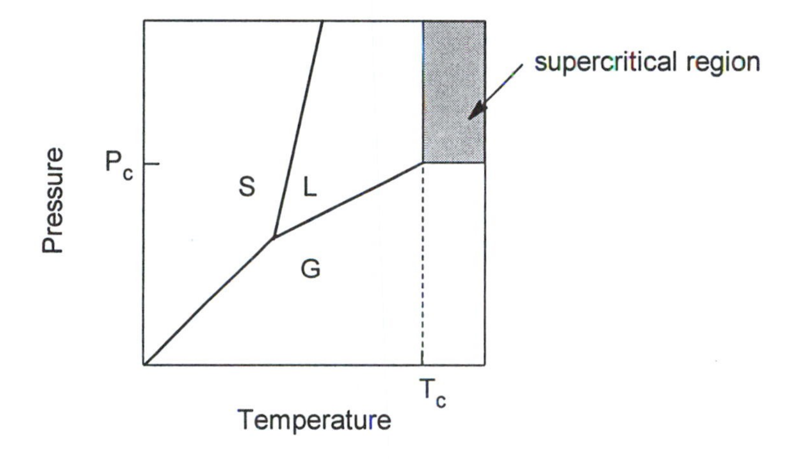

(a) Solid, liquid, and gas regions are labeled S, L, and G.

(b) Coexistence regions are represented by unbroken lines between pure phases.

(c) The critical isotherm appears as a broken vertical line at the point where $T = T_c$.

(d) The supercritical region is shown as a shaded area throughout which $T > T_c$ and $P > P_c$.

S11-16. Various key points are indicated on the phase diagrams that follow.

(a) The triple point marks the intersection of the solid–liquid, liquid-gas, and solid–gas coexistence lines. All three phases coexist in equilibrium at this particular combination of pressure and temperature:

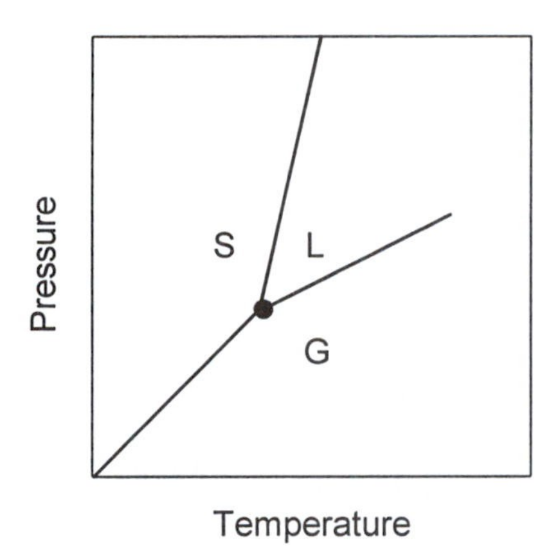

(b) The critical point terminates the liquid–gas coexistence line:

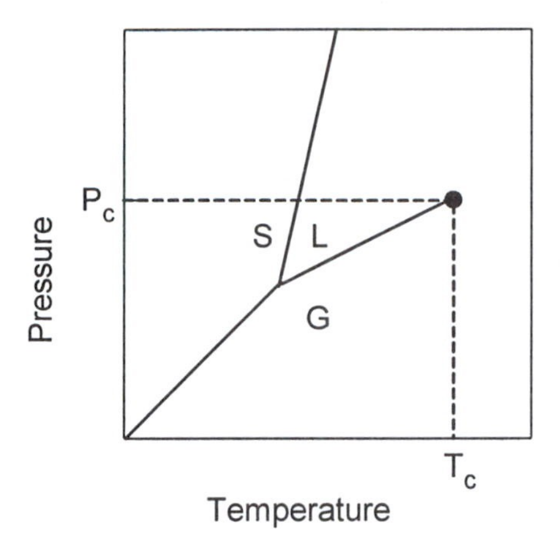

(c) The normal melting point marks the phase equilibrium between a liquid and solid at atmospheric pressure:

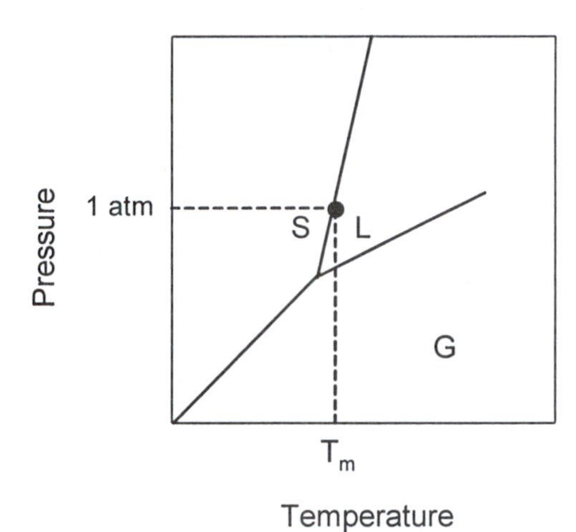

(d) The normal boiling point marks the phase equilibrium between a liquid and gas at atmospheric pressure:

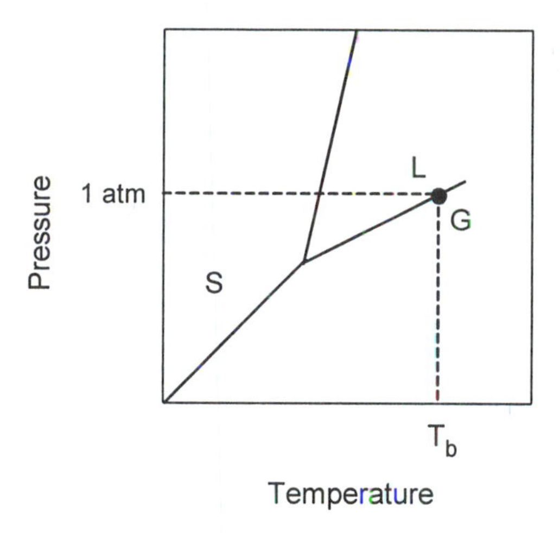

S11-17. Solid–liquid, liquid–gas, and solid–gas transitions are marked on the phase diagrams:

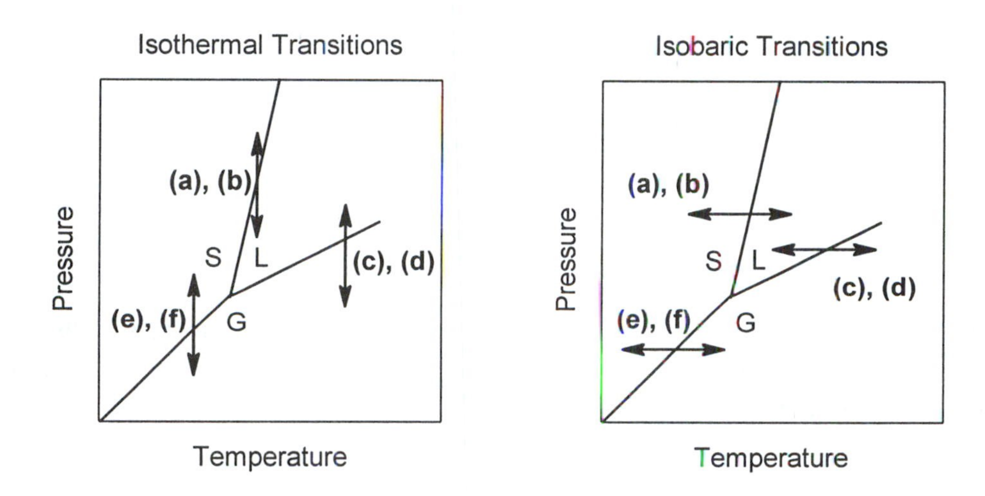

(a) Freezing: liquid → solid.

(b) Melting: solid → liquid.

(c) Condensation: gas → liquid.

(d) Vaporization: liquid → gas.

(e) Deposition: gas → solid.

(f) Sublimation: solid → gas.

S11-18. The solid–liquid line slopes upward from left to right:

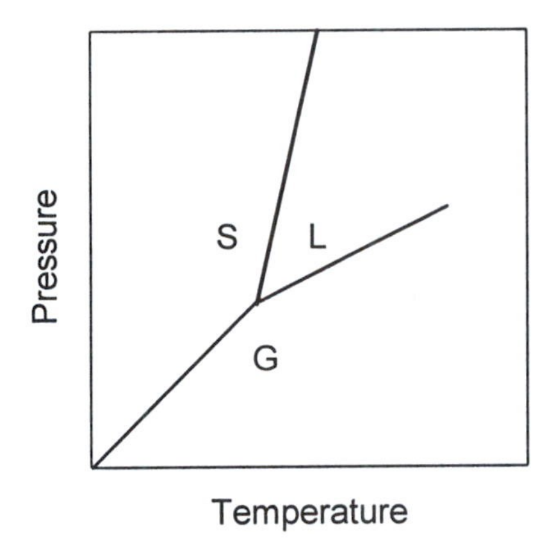

S11-19. The solid–liquid line has a negative slope:

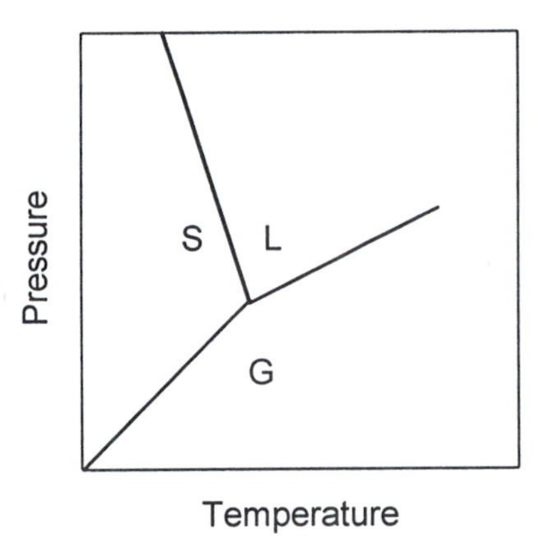

S11-20. The vapor pressure of substance 1 varies sharply with temperature, as evidenced by a steeply sloping liquid–gas coexistence line. The variation for substance 2, marked by a gentler slope, is less pronounced:

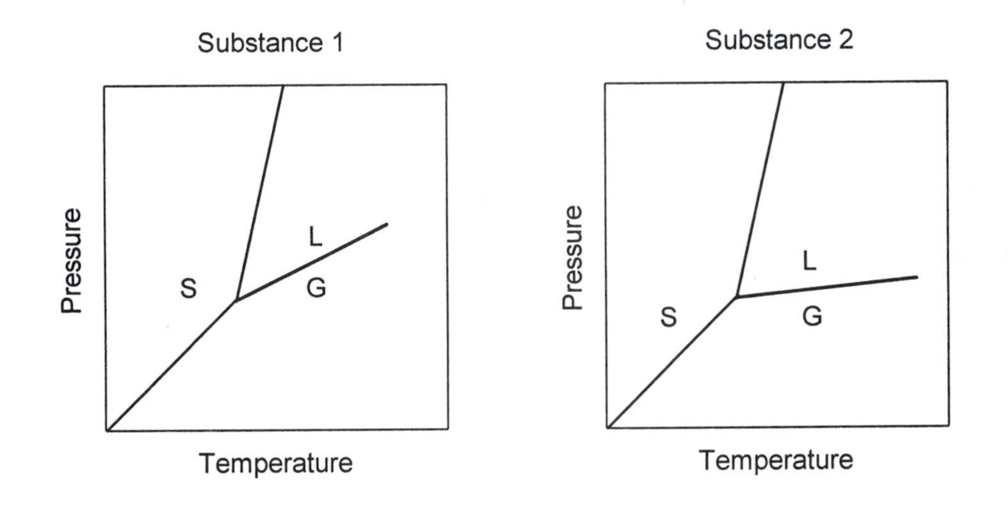

S11-21. The boiling temperature decreases from T_1 to T_2 as the ambient pressure decreases from P_1 to P_2:

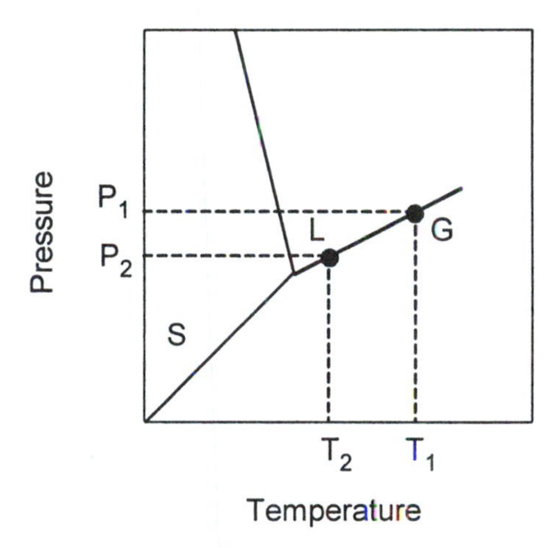

S11-22. Order increases in the direction gas → liquid → solid.

(a) The gas phase is disordered relative to the liquid phase.

(b) The liquid phase is disordered relative to the solid phase.

(c) The aqueous (solution) phase is disordered relative to the solid phase.

(d) The liquid phase at higher temperature (30°C) is disordered relative to the liquid phase at lower temperature (25°C).

Chapter 12

Equilibrium—The Stable State

S12-1. Classify each of the following situations as either a dynamic equilibrium or a static equilibrium:

(a) A framed picture hangs motionless on a wall.
(b) A constant vapor pressure exists above an enclosed liquid.
(c) All portions of a drink are equally sweet.
(d) The value of $[H_3O^+]$ is constant in an acidic solution.

S12-2. Consider the process

$$A(aq) + B(aq) \rightleftarrows C(aq) + D(g) \qquad K = 0.001$$

Propose a way to achieve 100% conversion of reactants into products, despite the relative smallness of the equilibrium constant.

S12-3. Set up, but do not solve, the algebraic expression needed to establish the equilibrium pressure of each component in the reaction below:

$$2A(g) + 3B(g) \rightleftarrows C(g) + 4D(g) \qquad K = 15$$

Assume the following values for the initial pressures:

$$P_{A,0} = 10 \text{ atm} \qquad P_{B,0} = 20 \text{ atm} \qquad P_{C,0} = P_{D,0} = 0$$

S12-4. Consider the process below:

$$A(g) + B(g) \rightleftarrows 3C(s) + D(g) \qquad K = 0.0001$$

After equilibrium is established, the volume of the system is suddenly tripled. Does this change in volume affect the equilibrium? If so, then in which direction does the reaction shift—toward products or toward reactants?

S12-5. Assume that $K = 2$ for the following process:

$$A(g) \rightleftarrows 2B(g)$$

(a) If $P_A = 1$ atm at equilibrium, what is the corresponding value of P_B? **(b)** Once this first equilibrium is attained, suppose that the volume of the system is halved and component B is removed completely. Calculate P_A and P_B at the new equilibrium.

S12-6. Say that the reaction

$$A(g) + B(g) \rightleftarrows C(g)$$

comes to equilibrium with final pressures as given below:

$$P_A = 0.50 \text{ atm} \qquad P_B = 0.50 \text{ atm} \qquad P_C = 1.0 \text{ atm}$$

(a) Calculate the equilibrium constant. **(b)** Three hours after the attainment of this first equilibrium, the product C is completely removed from the system. Calculate the values of P_A, P_B, and P_C after equilibrium is reestablished.

S12-7. Repeat part (b) of the preceding exercise, but with this one modification: The product C is removed from the system only 1 μs after the first equilibrium is reached, not 3 hours as specified earlier. Are the results affected in any way?

S12-8. Assume that P_D remains constant at 1 atm after the following reaction comes to equilibrium at the stated temperature:

$$A(g) + B(g) \rightleftarrows C(g) + D(g) \qquad K(500°C) = 100$$

Do you have sufficient information to determine the equilibrium pressures of all the other species? If yes, do so. If no, explain why the problem is poorly posed.

S12-9. Reconsider the hypothetical process introduced in the preceding exercise:

$$A(g) + B(g) \rightleftarrows C(g) + D(g) \qquad K(500°C) = 100$$

Given the initial pressures stated below,

$$P_{A,0} = P_{B,0} = 10 \text{ atm} \qquad P_{C,0} = P_{D,0} = 0$$

do you have sufficient information to determine the equilibrium pressures of all components? If yes, do so. If no, explain why the problem is poorly posed.

S12-10. Suppose that [A] = [B] = 1 *M* when the following reaction comes to equilibrium:

$$A(aq) \rightleftarrows B(aq) \qquad K = 1$$

Propose two different—but equally possible—sets of initial concentrations that will produce exactly the same concentrations at equilibrium.

S12-11. The equilibrium constant for the reaction

$$HF(aq) + H_2O(\ell) \rightleftarrows H_3O^+(aq) + F^-(aq)$$

is given by the expression

$$K = \frac{[H_3O^+][F^-]}{[HF]}$$

Assume that 5.00 grams of hydrofluoric acid are initially dissolved in 1.00 L of water at 25°C. If $K(25°C) = 6.8 \times 10^{-4}$, what are the concentrations of hydronium and fluoride ions at equilibrium?

S12-12. Repeat the preceding exercise, but this time use an expression that explicitly includes the concentration of water, a pure liquid:

$$K' = \frac{[H_3O^+][F^-]}{[HF][H_2O]}$$

Are the results changed in any way?

S12-13. The equilibrium constant for the reaction

$$Ag_2CO_3(s) \rightleftarrows 2Ag^+(aq) + CO_3^{2-}(aq)$$

is $K(25°C) = 8.5 \times 10^{-12}$. Calculate the equilibrium concentration of each ion in an aqueous solution of silver carbonate at 25°C.

S12-14. The equilibrium constant for the reaction

$$AgI(s) \rightleftarrows Ag^+(aq) + I^-(aq)$$

is $K(25°C) = 8.3 \times 10^{-17}$. **(a)** How many grams of silver iodide can be dissolved in 100 mL H_2O at 25°C? **(b)** How many grams can be dissolved in 100 L?

S12-15. Suppose that AgI is added to a 1.00 *M* solution of $AgNO_3$. At equilibrium, do you expect $[I^-]$ to be higher than, lower than, or the same as its value in a solution of AgI in pure water?

S12-16. Give the numerical value of K for the process

$$H_2O(g) \rightleftarrows H_2O(\ell)$$

at the following temperatures:

(a) 0°C **(b)** 25°C **(c)** 40°C **(d)** 80°C

Relevant information may be found in Appendix C.

S12-17. Does the equilibrium constant for the process

$$C_6H_6(\ell) \rightleftarrows C_6H_6(g)$$

increase, decrease, or remain the same as the temperature is increased?

S12-18. Does the equilibrium constant for the process

$$H_2O(g) \rightleftarrows H_2O(s)$$

increase, decrease, or remain the same as the temperature is increased?

S12-19. Suppose that the process

$$CH_3CH_2OH(\ell) \rightleftarrows CH_3CH_2OH(g)$$

has attained equilibrium. **(a)** Describe what will happen if the volume of the enclosed system is suddenly increased. **(b)** Does the equilibrium constant increase, decrease, or remain the same as the volume is increased? **(c)** What will happen if the system is left open to the atmosphere?

S12-20. The conversion of graphite into diamond,

$$C(s, \text{graphite}) \rightleftarrows C(s, \text{diamond})$$

proceeds on a geological time scale—a very long time. What can you say about the magnitude of K?

S12-21. The dissolution of sucrose in water,

$$C_{12}H_{22}O_{11}(s) \rightleftarrows C_{12}H_{22}O_{11}(aq)$$

proceeds in a matter of seconds. What can you say about the magnitude of K?

S12-22. A heap of NaCl remains undissolved in water until an experimenter vigorously stirs the solution. Does K increase, decrease, or remain the same as a result of the stirring?

S12-23. Our experimenter continues to dissolve NaCl in water until finally no more solid will dissolve, whereupon the additional sodium chloride falls to the bottom and resists even the most determined stirring. Does the position of equilibrium for the reaction

$$NaCl(s) \rightleftarrows Na^{+}(aq) + Cl^{-}(aq)$$

shift in any way as the excess solid accumulates? Is it legitimate to invoke Le Châteliers's principle?

SOLUTIONS

S12-1. A system in dynamic equilibrium is outwardly unchanging, inwardly in flux.

(a) *A framed picture hangs motionless on a wall.* Equilibrium arises from a static opposition of forces: gravity versus the tension in the wire.

(b) *A constant vapor pressure exists above an enclosed liquid.* The equilibrium is dynamic: molecules continually evaporate from the liquid and condense from the gas.

(c) *All portions of a drink are equally sweet.* Another dynamic equilibrium, brought about by an ongoing transport of matter.

(d) *The value of $[H_3O^+]$ is constant in an acidic solution.* An example of chemical equilibrium: the two-way, uninterrupted interconversion of reactants and products, by which a particular mass-action ratio is maintained. The equilibrium is dynamic.

S12-2. Drive the process to completion by allowing product D, a gas, to escape:

$$A(aq) + B(aq) \rightleftarrows C(aq) + D(g) \qquad K = 0.001$$

Removal of D pulls the reaction to the right, preventing the system from settling into equilibrium. The reaction quotient

$$Q = \frac{[\mathrm{C}]P_{\mathrm{D}}}{[\mathrm{A}][\mathrm{B}]}$$

remains less than K until the reactants are finally exhausted and the transformation ends.

S12-3. Given the initial pressures of A, B, C, and D, we use the balanced equation to express the pressures of all species at equilibrium. A minus sign indicates the disappearance of a reactant; a plus sign indicates the appearance of a product:

$$2\mathrm{A(g)} + 3\mathrm{B(g)} \rightleftarrows \mathrm{C(g)} + 4\mathrm{D(g)} \qquad K = 15$$

	2A(g)	3B(g)	C(g)	4D(g)
Initial pressure	10	20	0	0
Change	$-2x$	$-3x$	$+x$	$+4x$
Equilibrium pressure	$10 - 2x$	$20 - 3x$	x	$4x$

The equilibrium equation is prescribed by the law of mass action:

$$K = \frac{P_{\mathrm{C}} P_{\mathrm{D}}^4}{P_{\mathrm{A}}^2 P_{\mathrm{B}}^3} = \frac{x(4x)^4}{(10-2x)^2(20-3x)^3} = 15$$

S12-4. Two moles of gas on the left are in equilibrium with one mole of gas on the right:

$$\mathrm{A(g)} + \mathrm{B(g)} \rightleftarrows 3\mathrm{C(s)} + \mathrm{D(g)}$$

$$K = \frac{P_{\mathrm{D}}}{P_{\mathrm{A}} P_{\mathrm{B}}} = 0.0001$$

The concentration of C, a solid, does not appear in the equilibrium constant.

If we now triple the volume, then each equilibrium partial pressure

$$P_i = \frac{n_i RT}{V}$$

is reduced abruptly by a factor of 3:

$$P_{i,\,\mathrm{noneq}} = \frac{n_i RT}{3V} = \frac{P_i}{3}$$

The increase in volume leaves the reaction quotient Q greater than the equilibrium constant K,

$$Q = \frac{\frac{1}{3}P_D}{\left(\frac{1}{3}P_A\right)\left(\frac{1}{3}P_B\right)} = 3\frac{P_D}{P_A P_B} = 3K$$

and the transformation shifts to the left, toward reactants. The numerical value of K is itself unchanged.

S12-5. We consider the following gas-phase process:

$$A(g) \rightleftarrows 2B(g)$$

$$K = \frac{P_B^2}{P_A} = 2$$

(a) The equilibrium pressures of A and B are constrained by the value of K:

$$P_B = \sqrt{P_A K} = \sqrt{1 \times 2} = \sqrt{2} \text{ atm}$$

(b) Each partial pressure doubles when the volume is abruptly halved,

$$P_{A,eq} = 1 \text{ atm} \qquad P_{A,noneq} = 2 \text{ atm}$$

$$P_{B,eq} = \sqrt{2} \text{ atm} \qquad P_{B,noneq} = 2\sqrt{2} \text{ atm}$$

but the problem stipulates that all of component B is subsequently withdrawn. Hence the initial nonequilibrium pressure of B is actually zero:

	A(g)	⇄ 2B(g)
Initial pressure	2	0
Change	$-x$	$2x$
Equilibrium pressure	$2-x$	$2x$

We now have enough information to set up and solve the mass-action equation:

$$2 = \frac{P_B^2}{P_A} = \frac{(2x)^2}{2-x}$$

$$0 = 2x^2 + x - 2$$

$$x = \frac{-1 \pm \sqrt{(1)^2 - 4(2)(-2)}}{2(2)} = 0.781, \ -1.281$$

The physically valid solution is $x = 0.781$:

$$P_A = 2 - x = 1.219 \text{ atm}$$

$$P_B = 2x = 1.562 \text{ atm}$$

S12-6. We consider a hypothetical gas-phase reaction:

$$A(g) + B(g) \rightleftarrows C(g)$$

$$K = \frac{P_C}{P_A P_B}$$

(a) Given a set of equilibrium pressures, we can evaluate K directly:

$$K = \frac{P_C}{P_A P_B} = \frac{1.0}{(0.50)(0.50)} = 4.0$$

(b) Initial and equilibrium pressures are shown symbolically below:

	A(g)	+	B(g)	$\rightleftarrows$	C(g)
Initial pressure	0.50		0.50		0
Change	$-x$		$-x$		x
Equilibrium pressure	$0.50 - x$		$0.50 - x$		x

The resulting quadratic equation is easily factored:

$$4.0 = \frac{P_C}{P_A P_B} = \frac{x}{(0.50 - x)^2}$$

$$0 = 4x^2 - 5x + 1$$

$$0 = (4x - 1)(x - 1)$$

$$x = 0.25,\ 1.0$$

We take $x = 0.25$ as the physically valid solution:

$$P_A = P_B = 0.50 - x = 0.25 \text{ atm}$$

$$P_C = x = 0.25 \text{ atm}$$

S12-7. The results are the same as in the preceding exercise. Macroscopic equilibrium concentrations remain constant, independent of time, for as long as the equilibrium persists.

S12-8. The information given is insufficient. We need to know the initial pressures of all species *before* the system equilibrates. Otherwise, at equilibrium we are faced with a single equation in three unknowns:

$$\begin{array}{ccccccc} A(g) & + & B(g) & \rightleftarrows & C(g) & + & D(g) \\ x & & y & & z & & 1 \end{array}$$

$$K = \frac{P_C P_D}{P_A P_B} = \frac{z(1)}{xy} = 100$$

Once equilibrium is attained, all memory of the initial conditions is lost.

S12-9. Given the initial pressure of each species, we now have enough information to determine the equilibrium concentrations:

$$A(g) + B(g) \rightleftarrows C(g) + D(g)$$

	A(g)	B(g)	C(g)	D(g)
Initial pressure	10	10	0	0
Change	$-x$	$-x$	x	x
Equilibrium pressure	$10-x$	$10-x$	x	x

The mass-action expression yields a quadratic equation:

$$K = \frac{P_C P_D}{P_A P_B}$$

$$100 = \frac{x^2}{(10-x)^2}$$

$$0 = 99x^2 - 2000x + 10{,}000$$

$$x = \frac{-(-2000) \pm \sqrt{(-2000)^2 - 4(99)(10{,}000)}}{2(99)} = 11.11,\ \ 9.09$$

The physically valid solution is $x = 9.09$:

$$P_A = P_B = 10 - x = 0.91 \text{ atm}$$

$$P_C = P_D = x = 9.09 \text{ atm}$$

S12-10. Both sets of initial concentrations below will produce $[A] = [B] = 1\ M$ when the reaction $A \rightleftarrows B$ comes to equilibrium:

$[A]_0 = 2\ M,\ \ [B]_0 = 0$ and $[A]_0 = 0,\ \ [B]_0 = 2\ M$

The equilibrium equation is trivial in each case:

$$\begin{array}{ccc} \mathrm{A(aq)} & \rightleftarrows & \mathrm{B(aq)} \\ 2-x & & x \end{array} \qquad\qquad \begin{array}{ccc} \mathrm{A(aq)} & \rightleftarrows & \mathrm{B(aq)} \\ x & & 2-x \end{array}$$

$$K = \frac{[\mathrm{B}]}{[\mathrm{A}]} = \frac{x}{2-x} = 1 \qquad\qquad K = \frac{[\mathrm{B}]}{[\mathrm{A}]} = \frac{2-x}{x} = 1$$

$$x = 2 - x \qquad\qquad x = 2 - x$$

$$x = 1 \qquad\qquad x = 1$$

$$[\mathrm{A}] = [\mathrm{B}] = 1\ M$$

In general, we arrive at the same final mass-action ratio—unity—by starting from an arbitrary set of initial concentrations c_A and c_B:

$$\begin{array}{ccc} \mathrm{A(aq)} & \rightleftarrows & \mathrm{B(aq)} \\ c_A - x & & c_B + x \end{array} \qquad\qquad (c_A > c_B)$$

$$K = \frac{[\mathrm{B}]}{[\mathrm{A}]} = \frac{c_B + x}{c_A - x} = 1$$

$$c_B + x = c_A - x$$

$$x = \frac{c_A - c_B}{2}$$

S12-11. Beginning with the initial concentration of hydrofluoric acid,

$$[\mathrm{HF}]_0 = \frac{\left(5.00\ \mathrm{g\ HF} \times \dfrac{1\ \mathrm{mol\ HF}}{20.0063\ \mathrm{g\ HF}}\right)}{1.00\ \mathrm{L}} = 0.250\ M$$

we proceed to set up the equilibrium expression:

$$\begin{array}{ccccccc} \mathrm{HF(aq)} & + & \mathrm{H_2O}(\ell) & \rightleftarrows & \mathrm{H_3O^+(aq)} & + & \mathrm{F^-(aq)} \\ 0.250 - x & & & & x & & x \end{array}$$

$$K = \frac{[\mathrm{H_3O^+}][\mathrm{F^-}]}{[\mathrm{HF}]} = \frac{x^2}{0.250 - x} = 6.8 \times 10^{-4}$$

The equation reduces to a simple square root, since $0.250 - x \approx 0.250$ (to within 5%):

$$\frac{x^2}{0.250} \approx 6.8 \times 10^{-4} \quad \text{if} \quad x << 0.250$$

$$x^2 = 1.7 \times 10^{-4}$$

$$x = 0.013$$

The final concentrations of H_3O^+ and F^- are both 0.013 *M*.

S12-12. The concentration of liquid water remains effectively constant throughout the reaction, largely undiminished by the weak ionization of HF:

$$[H_2O(\ell)] = \frac{1000 \text{ g } H_2O(\ell) \times \dfrac{1 \text{ mol } H_2O}{18.015 \text{ g } H_2O}}{\text{L}} = 55.5\ M$$

As a result, we can define and evaluate a modified equilibrium constant K':

$$K' = \frac{1}{[H_2O(\ell)]} \times \frac{[H_3O^+][F^-]}{[HF]} = \frac{K}{[H_2O(\ell)]} = \frac{K}{55.5}$$

The equilibrium calculation is otherwise unchanged:

$$\begin{array}{ccccccc} HF(aq) & + & H_2O(\ell) & \rightleftarrows & H_3O^+(aq) & + & F^-(aq) \\ 0.250 - x & & 55.5 & & x & & x \end{array}$$

$$\frac{[H_3O^+][F^-]}{[HF][H_2O]} = K' = \frac{K}{[H_2O]}$$

$$\frac{x^2}{(0.250 - x)(55.5)} = \frac{6.8 \times 10^{-4}}{55.5}$$

$$\frac{x^2}{0.250 - x} = 6.8 \times 10^{-4}$$

We solve exactly the same equation as before.

S12-13. Set up and solve the mass-action expression:

$$Ag_2CO_3(s) \rightleftarrows 2Ag^+(aq) + CO_3^{2-}(aq)$$

$$\qquad 2x \qquad\qquad x$$

$$[Ag^+]^2[CO_3^{2-}] = K$$

$$(2x)^2 x = 8.5 \times 10^{-12}$$

$$x = 1.3 \times 10^{-4}$$

The equilibrium concentration of each ion is given below:

$$[Ag^+] = 2x = 2.6 \times 10^{-4}\ M$$

$$[CO_3^{2-}] = x = 1.3 \times 10^{-4}\ M$$

S12-14. The equilibrium concentration of AgI in pure water is equal to 9.1×10^{-9} *M*:

$$AgI(s) \rightleftarrows Ag^+(aq) + I^-(aq)$$

$$\qquad x \qquad\qquad x$$

$$[Ag^+][I^-] = K$$

$$x^2 = 8.3 \times 10^{-17}$$

$$x = 9.1 \times 10^{-9}$$

(a) Use the molar mass and specified volume to obtain a mass in grams:

$$\frac{9.1 \times 10^{-9}\text{ mol AgI}}{\text{L}} \times \frac{234.773\text{ g AgI}}{\text{mol AgI}} \times 0.100\text{ L} = 2.1 \times 10^{-7}\text{ g AgI}$$

(b) Similar:

$$\frac{9.1 \times 10^{-9}\text{ mol AgI}}{\text{L}} \times \frac{234.773\text{ g AgI}}{\text{mol AgI}} \times 100\text{ L} = 2.1 \times 10^{-4}\text{ g AgI}$$

S12-15. *Lower*. The large existing concentration of Ag^+ in aqueous $AgNO_3$ (1.00 *M*) suppresses the subsequent dissolution of I^-:

$$\mathrm{AgI(s)} \rightleftarrows \mathrm{Ag^+(aq)} + \mathrm{I^-(aq)}$$
$$\qquad\qquad 1.00 + x \qquad x$$

$$[\mathrm{Ag^+}][\mathrm{I^-}] = K$$

$$(1.00 + x)x = 8.3 \times 10^{-17}$$

$$x \approx 8.3 \times 10^{-17}$$

S12-16. Consider the condensation of water vapor:

$$\mathrm{H_2O(g)} \rightleftarrows \mathrm{H_2O(\ell)}$$

$$K(T) = \frac{1}{P_{\mathrm{H_2O}}(T)}$$

To compute a value for *K*(*T*), we look up the vapor pressure of water at the appropriate temperature (*Principles of Chemistry*, Table C-15).

For example, $P_{\mathrm{H_2O}}$ is 4.6 torr at 0°C. We convert torr into atmospheres,

$$4.6 \text{ torr} \times \frac{1 \text{ atm}}{760 \text{ torr}} = 0.00605 \text{ atm}$$

and insert the pressure into the expression for *K*:

$$K(0^\circ\,\mathrm{C}) = \frac{1}{P_{\mathrm{H_2O}}(0^\circ\,\mathrm{C})} = \frac{1}{0.00605} = 1.7 \times 10^2$$

Results are summarized below:

	T (°C)	$P_{\mathrm{H_2O}}$ (torr)	*K*
(a)	0	4.6	1.7×10^2
(b)	25	23.8	31.9
(c)	40	55.4	13.7
(d)	80	355.3	2.139

S12-17. The vapor pressure of a liquid increases as the temperature is increased. Hence the equilibrium constant

$$K = P_{C_6H_6}$$

for the endothermic process

$$\text{heat} + C_6H_6(\ell) \rightleftarrows C_6H_6(g)$$

also increases with temperature. Heat functions as a reactant in the conversion of a liquid phase into a chaotically disordered gas phase.

S12-18. The equilibrium constant

$$K = \frac{1}{P_{H_2O}}$$

for the exothermic process

$$H_2O(g) \rightleftarrows H_2O(s) + \text{heat}$$

decreases as the temperature increases. The gaseous form (on the left) is favored at high temperature.

S12-19. The process

$$CH_3CH_2OH(\ell) \rightleftarrows CH_3CH_2OH(g)$$

can be written in general form as

$$A(\ell) \rightleftarrows A(g)$$

with

$$K = P_A$$

(a) If the volume is increased, then P_A will decrease and the reaction quotient Q will fall to a value less than K. The process shifts to the right, toward the gas, as the system moves to restore equilibrium. More gas is produced at the expense of liquid.

(b) The equilibrium constant is unchanged. Both its form and numerical value remain unaffected by any alteration in volume.

(c) If the system is left open to the atmosphere, then the gaseous product will escape. Unable to establish equilibrium, the reaction will be driven to completion.

S12-20. The elapsed time of reaction, a kinetic consideration, has no bearing on the position of equilibrium. We can say nothing about the value of K.

S12-21. Again, the speed of reaction is irrelevant to the mass-action expression. We can say nothing about the value of K.

S12-22. The equilibrium constant is unaffected by the stirring, provided that the temperature remains unchanged. Stirring affects only the kinetics of the reaction; the same amount of material eventually dissolves, although faster.

S12-23. The position of equilibrium for the reaction

$$\mathrm{NaCl(s)} \rightleftarrows \mathrm{Na^+(aq)} + \mathrm{Cl^-(aq)}$$

is determined by a mass-action expression that does not include solid NaCl:

$$K = [\mathrm{Na^+}][\mathrm{Cl^-}]$$

So long as *some* undissolved NaCl is on hand to maintain the heterogeneous equilibrium, the actual amount of solid is irrelevant. The accumulation of undissolved solute does not constitute a stress on the system in the sense required by Le Châtelier's principle.

Chapter 13

Energy, Heat, and Chemical Change

S13-1. Classify each property as extensive or intensive:

(a) thermal conductivity **(b)** melting point **(c)** triple point **(d)** color

Assume that the system is sufficiently large to behave in bulk.

S13-2. Which of the following quantities are functions of state? Which are functions of path?

(a) $H - PV$ **(b)** $E - TS$ **(c)** $c_V \Delta T$ **(d)** $\frac{3}{2}R \Delta T$

S13-3. A fixed quantity of gaseous CO_2 (1 mol), initially at STP, is subjected to a rapid cycle of expansion and compression. When the process is over, the temperature is 273 K and the volume is 22.4 L. Assume that the gas behaves ideally. **(a)** Calculate ΔE. **(b)** Calculate ΔH. **(c)** Calculate ΔS.

S13-4. How much work is done when an ideal gas expands from 1.00 L to 3.00 L under a constant external pressure of 2.00 atm?

S13-5. Suppose that an ideal gas is confined to an insulating vessel, unable to exchange heat with its surroundings. Assume further that the internal pressure is equal to the external pressure. By itself, unassisted, is the gas able to expand?

S13-6. A sample of helium gas at STP is confined by a piston to a volume of 10.0 L. System and surroundings are both at the same temperature, able to exchange heat freely. **(a)** When the pressure is doubled, the volume of gas falls to 5.0 L. Calculate ΔE. **(b)** Is there a flow of heat? If so, in which direction does it proceed—from system to surroundings, or from surroundings to system?

S13-7. Consider the following two states of an ideal gas, both at STP:

State 1: $n = 1.00$ mol State 2: $n = 2.00$ mol

(a) Calculate $\Delta E = E_2 - E_1$. **(b)** Calculate $\Delta H = H_2 - H_1$.

S13-8. The temperature of argon gas in a closed system (1.00 mol) falls from 300 K to 200 K at constant volume. **(a)** In which direction does the heat flow? **(b)** Do you have enough information to compute w? If yes, do so. If no, explain why. **(c)** Do you have enough information to compute q, ΔE, and ΔH? If yes, do so. If no, explain why.

S13-9. A system of neon gas at STP is confined to a rigid volume of 224.1 L. Calculate ΔE and ΔT after the gas absorbs a heat flow of 10.0 kJ.

S13-10. The temperature of helium gas in a rigid vessel rises by 10.0 K after the system absorbs 5.00 kJ of heat. Calculate the mass of helium in the vessel.

S13-11. A beaker of water containing 1.00 mol, open to the atmosphere at 20°C, is brought into contact with an object at a lower temperature. If the temperature of the water falls by 10.15°C, what is the corresponding change in enthalpy? Relevant data may be found in Appendix C.

S13-12. Consult Appendix C for any data required to answer the following questions. **(a)** How many joules are needed to raise the temperature of 3.17 grams of potassium metal by 15.0°C at constant pressure? **(b)** How many joules are needed to bring about the same change in temperature for 3.17 grams of lithium?

S13-13. What additional information about the system do you need in order to compute $\Delta H°$ for the indicated reaction?

$$2H_2 + O_2 \rightarrow 2H_2O$$

S13-14. Use the data in Appendix C to calculate $\Delta H°$ for each of the reactions below:

(a) $4NH_3(g) + 5O_2(g) \rightarrow 4NO(g) + 6H_2O(g)$
(b) $4NH_3(g) + 3O_2(g) \rightarrow 2N_2(g) + 6H_2O(g)$
(c) $2NH_3(g) + 2O_2(g) \rightarrow N_2O(g) + 3H_2O(g)$

S13-15. Use the data in Appendix C to calculate the heat of combustion per mole of liquid octane at 1 atm and 25°C.

S13-16. How much heat must be supplied to vaporize 100.0 grams of metallic silver at 1 atm and 25°C? Consult Appendix C as needed.

S13-17. Again, use Appendix C where needed. **(a)** Calculate the change in enthalpy when a 456-g sample of benzene condenses from gas to liquid at 1 atm and 25°C. **(b)** Assume that the heat evolved is captured entirely by 5.00 L of water at 25.00°C. What is the final temperature of the water?

S13-18. Similar. **(a)** Calculate the change in enthalpy when a 755-mL sample of ethanol vaporizes at 1 atm and 25°C. The density of ethanol is 0.7873 g mL^{-1}. **(b)** Assume that the heat required for this vaporization is withdrawn from 1000 moles of helium at 25.0°C, the transfer occurring at constant pressure. What is the final temperature of the helium?

S13-19. Suppose that a typographical error has caused all the enthalpies of formation in Table C-16 to be exactly 1.0 kJ mol^{-1} higher than intended. Would our calculations be affected in any way? If so, how?

S13-20. How can we say that heat (q) is not a function of state, but that enthalpy ($H \equiv q_P$) *is* a function of state? After all, isn't enthalpy equivalent to the heat exchanged under constant pressure?

S13-21. The dissolution of potassium chloride in water is an endothermic reaction. Would you expect a beaker of water to grow warm to the touch or cold to the touch as KCl is stirred into solution?

S13-22. Say that the dissolution of a certain salt in water is an exothermic reaction. Would you expect more material or less material to dissolve as the temperature is increased?

S13-23. Show, algebraically, that ΔH for the reaction

$$a\text{A} + b\text{B} \rightarrow c\text{C} + d\text{D}$$

is equal in magnitude and opposite in sign to ΔH for the reaction

$$c\text{C} + d\text{D} \rightarrow a\text{A} + b\text{B}$$

S13-24. Calculate the enthalpy of sublimation for iodine at 1 atm and 25°C.

S13-25. Consider the following process:

$$2\text{H(g)} \rightarrow \text{H}_2\text{(g)} \qquad \Delta H^\circ = -436 \text{ kJ}$$

What is the standard heat of formation for $H_2(g)$? Does the reaction enthalpy specified above imply somehow that $\Delta H_f^\circ = -436$ kJ mol^{-1} for molecular hydrogen—in contradiction to our convention concerning elements in their standard states?

SOLUTIONS

S13-1. An *intensive* property, independent of the amount of material, has a value at every point in a sample. An *extensive* property varies with the amount of material and requires the entire system for its definition.

In the assessments below we assume that the system is homogeneous and behaves in bulk, deriving its macroscopic properties from a statistically large number of particles. Surface and cluster effects, for example, are not considered.

(a) Thermal conductivity is an intensive property, its value the same throughout the system.

(b) The melting point of a pure substance is an intensive property. Its value is the same in a homogeneous system of any size.

(c) Similarly, the triple point of a pure substance is also an intensive property.

(d) Color, although a complex phenomenon, usually presents itself as an intensive property—recognizable at every point in a sample.

S13-2. A *function of state* is determined only by the current equilibrium condition of a system, as specified by a set of state variables. A *function of path* varies with the history and preparation of a system.

(a) The quantity $H - PV$ is a function of state, an alternative expression of the internal energy:

$$E = H - PV$$

(b) The quantity $E - TS$ is also a function of state, formed by combination of the state functions E and S.

(c) The quantity $c_V \Delta T$ is equivalent to a flow of *heat*, a function of path in systems where $\Delta V \neq 0$.

(d) The quantity $\frac{3}{2}R\,\Delta T$, the molar energy of an ideal gas, is a function of state.

S13-3. The initial and final states are the same, as demanded by the equation $PV = nRT$:

$$P_1 = 1 \text{ atm} \qquad V_1 = 22.4 \text{ L} \qquad T_1 = 273 \text{ K} \qquad n_1 = 1 \text{ mol}$$

$$P_2 = 1 \text{ atm} \qquad V_2 = 22.4 \text{ L} \qquad T_2 = 273 \text{ K} \qquad n_2 = 1 \text{ mol}$$

All state functions, among them E, H, and S, are unchanged as a result.

(a) $\Delta E = 0$

(b) $\Delta H = 0$

(c) $\Delta S = 0$

S13-4. The system does an amount of work equal to –405 J:

$$w = -P_{\text{ext}}(V_2 - V_1) = -(2.00 \text{ atm})(3.00 \text{ L} - 1.00 \text{ L}) \times \frac{8.3145 \text{ J}}{0.082058 \text{ atm L}} = -405 \text{ J}$$

S13-5. The system cannot draw heat from its surroundings to do the necessary work. It is unable to expand.

Under the conditions specified, an expansion can be effected only by a reduction in external pressure or an increase in the amount of material inside the system.

S13-6. The initial state of the ideal gas is specified as follows:

$$P_1 = 1.00 \text{ atm} \qquad V_1 = 10.0 \text{ L} \qquad T_1 = 273 \text{ K}$$

With the amount of material fixed, the internal energy depends only on temperature:

$$\Delta E = \frac{3}{2} nR\,\Delta T$$

(a) The change in internal energy is zero, since there is no change in temperature:

$$\frac{P_1V_1}{T_1} = \frac{P_2V_2}{T_2}$$

$$T_2 = \frac{P_2}{P_1} \times \frac{V_2}{V_1} \times T_1 = \frac{2.00 \text{ atm}}{1.00 \text{ atm}} \times \frac{5.0 \text{ L}}{10.0 \text{ L}} \times T_1 = T_1$$

(b) Work is done *on* the gas to compress it. An equal amount of heat flows from the gas (the system) to the surroundings outside:

$$\Delta E = q + w = 0$$

$$q = -w$$

S13-7. Energy and enthalpy are extensive properties, directly proportional to the amount of material.

(a) The internal energy doubles when n is doubled at constant temperature:

$$\Delta E = E_2 - E_1 = \frac{3}{2}RT(n_2 - n_1) = \frac{3}{2}(8.3145 \text{ J mol}^{-1} \text{ K}^{-1})(273.15 \text{ K})(2.00 \text{ mol} - 1.00 \text{ mol})$$

$$= 3.41 \times 10^3 \text{ J}$$

(b) Use $PV = nRT$ to state the enthalpy in terms of n and T:

$$\Delta H = \Delta E + \Delta(PV) = \Delta E + \Delta(nRT)$$

The quantities ΔH and ΔE thus differ by RT under the conditions prescribed:

$$\Delta H = \Delta E + RT\,\Delta n = 3.41 \times 10^3 \text{ J} + (8.3145 \text{ J mol}^{-1} \text{ K}^{-1})(273.15 \text{ K})(2.00 \text{ mol} - 1.00 \text{ mol})$$

$$= 5.68 \times 10^3 \text{ J}$$

S13-8. We are given the following information about the initial and final states:

$$V_1 = V_0 \qquad n_1 = 1.00 \text{ mol} \qquad T_1 = 300 \text{ K}$$

$$V_2 = V_0 \qquad n_2 = 1.00 \text{ mol} \qquad T_2 = 200 \text{ K}$$

(a) Heat flows out of the gas, from system to surroundings. The temperature of the system decreases.

(b) The work done is *zero*, since the process occurs at constant volume:

$$w = -P_{\text{ext}}\,\Delta V = 0$$

(c) Knowledge of ΔT enables us to calculate ΔE, q, and ΔH for the ideal gas.

If the amount of material is fixed, then the change in internal energy is directly proportional to the change in temperature:

$$\Delta E = \frac{3}{2}nR\,\Delta T = \frac{3}{2}(1.00 \text{ mol})(8.3145 \text{ J mol}^{-1} \text{ K}^{-1})(200 \text{ K} - 300 \text{ K}) = -1.25 \times 10^3 \text{ J}$$

The accompanying flow of heat is constrained by the first law of thermodynamics:

$$\Delta E = q + w = q + 0 = q$$

$$q = \Delta E = -1.25 \times 10^3 \text{ J}$$

The change in enthalpy, finally, differs from ΔE by a term $nR\,\Delta T$, equivalent here to $\Delta(PV)$:

$$\Delta H = \Delta E + nR\,\Delta T = -1.25 \times 10^3 \text{ J} + (1.00 \text{ mol})(8.3145 \text{ J mol}^{-1} \text{ K}^{-1})(200 \text{ K} - 300 \text{ K})$$

$$= -2.08 \times 10^3 \text{ J}$$

S13-9. No work is done, because the transformation occurs at constant volume. The change in internal energy thus becomes equal to the heat absorbed:

$$\Delta E = q + w = q + 0 = q = 10.0 \text{ kJ}$$

We know, moreover, that our present system ($V = 224.1$ L) contains 10.00 moles of gas, given that one mole of gas occupies a volume of 22.41 L at STP. The increase in temperature is directly proportional to the change in energy:

$$\Delta E = \frac{3}{2} nR\,\Delta T$$

$$\Delta T = \frac{2\,\Delta E}{3nR} = \frac{2(10.0 \times 10^3 \text{ J})}{3(10.00 \text{ mol})(8.3145 \text{ J mol}^{-1} \text{ K}^{-1})} = 80.2 \text{ K}$$

S13-10. This problem is similar to the two preceding. For variety, we explicitly identify the molar heat capacity at constant volume ($c_V = \frac{3}{2}R$) while doing the calculation.

The symbols $\mathcal{m}$ and m denote, respectively, the molar mass and gram mass of helium:

$$q_V = nc_V\,\Delta T = \left(\frac{m}{\mathcal{m}}\right) c_V\,\Delta T = \left(\frac{m}{\mathcal{m}}\right)\left(\frac{3}{2}R\right) \Delta T$$

$$m = \frac{2\mathcal{m}\,q_V}{3R\,\Delta T} = \frac{2(4.003 \text{ g mol}^{-1})(5.00 \times 10^3 \text{ J})}{3(8.3145 \text{ J mol}^{-1} \text{ K}^{-1})(10.0 \text{ K})} = 160. \text{ g}$$

S13-11. Note that the process occurs at constant pressure:

$$\Delta H = q_P = nc_P\,\Delta T$$

From Table C-15 in *Principles of Chemistry* we obtain the value

$$c_P = 75.34 \text{ J mol}^{-1} \text{ K}^{-1}$$

at 20°C. Calculation of ΔH then follows straightforwardly:

$$\Delta H = (1.00 \text{ mol})(75.34 \text{ J mol}^{-1} \text{ K}^{-1})(-10.15 \text{ K}) = -765 \text{ J}$$

S13-12. Molar heat capacity and specific heat capacity.

(a) From Table C-16 in *Principles of Chemistry* we obtain the value

$$c_P = 29.6 \text{ J mol}^{-1} \text{ K}^{-1}$$

for the molar heat capacity of metallic potassium (at constant pressure). The calculation of heat flow is carried out in the usual way:

$$q_P = nc_P\,\Delta T = \left(3.17 \text{ g} \times \frac{1 \text{ mol}}{39.0983 \text{ g}}\right)\left(\frac{29.6 \text{ J}}{\text{mol K}}\right)(15.0 \text{ K}) = 36.0 \text{ J}$$

(b) Using the value $c_P = 24.8 \text{ J mol}^{-1} \text{ K}^{-1}$ for metallic lithium, we establish that $q_P = 170.$ J:

$$q_P = nc_P\,\Delta T = \left(3.17 \text{ g} \times \frac{1 \text{ mol}}{6.941 \text{ g}}\right)\left(\frac{24.8 \text{ J}}{\text{mol K}}\right)(15.0 \text{ K}) = 170. \text{ J}$$

Doing so, we note that the molar mass individually determines the specific heat of each material (heat capacity per gram).

S13-13. We need to know the state of matter—solid, liquid, or gas—for each of the species involved in the transformation.

S13-14. Hess's law, reflecting simply the conservation of energy, enables us to calculate the standard change in enthalpy for an arbitrary chemical reaction:

$$a\text{A} + b\text{B} \rightarrow c\text{C} + d\text{D}$$

$$\Delta H^\circ = c\,\Delta H_f^\circ(\text{C}) + d\,\Delta H_f^\circ(\text{D}) - a\,\Delta H_f^\circ(\text{A}) - b\,\Delta H_f^\circ(\text{B})$$

The standard enthalpy of formation, ΔH_f°, is zero for any element in its standard state under standard conditions. See Table C-16 in Appendix C of *Principles of Chemistry* (pages A85–A92).

(a) The combustion of ammonia to nitric oxide and water is strongly exothermic:

$$4NH_3(g) + 5O_2(g) \rightarrow 4NO(g) + 6H_2O(g)$$

$$\Delta H^\circ = 4\,\Delta H_f^\circ[\text{NO(g)}] + 6\,\Delta H_f^\circ[\text{H}_2\text{O(g)}] - 4\,\Delta H_f^\circ[\text{NH}_3\text{(g)}] - 5\,\Delta H_f^\circ[\text{O}_2\text{(g)}]$$

$$= \left(\frac{90.3 \text{ kJ}}{\text{mol}} \times 4 \text{ mol}\right) + \left(-\frac{241.8 \text{ kJ}}{\text{mol}} \times 6 \text{ mol}\right) - \left(-\frac{46.1 \text{ kJ}}{\text{mol}} \times 4 \text{ mol}\right) - \left(\frac{0 \text{ kJ}}{\text{mol}} \times 5 \text{ mol}\right)$$

$$= -905.2 \text{ kJ}$$

(b) The same reactants, ammonia and oxygen, combine to yield a different set of products—and release a different quantity of heat as well:

$$4NH_3(g) + 3O_2(g) \rightarrow 2N_2(g) + 6H_2O(g)$$

$$\Delta H^\circ = 2\,\Delta H_f^\circ[N_2(g)] + 6\,\Delta H_f^\circ[H_2O(g)] - 4\,\Delta H_f^\circ[NH_3(g)] - 3\,\Delta H_f^\circ[O_2(g)]$$

$$= \left(\frac{0\text{ kJ}}{\text{mol}} \times 2\text{ mol}\right) + \left(-\frac{241.8\text{ kJ}}{\text{mol}} \times 6\text{ mol}\right) - \left(-\frac{46.1\text{ kJ}}{\text{mol}} \times 4\text{ mol}\right) - \left(\frac{0\text{ kJ}}{\text{mol}} \times 3\text{ mol}\right)$$

$$= -1266.4\text{ kJ}$$

(c) Another reaction of ammonia and oxygen, and another value of ΔH°:

$$2NH_3(g) + 2O_2(g) \rightarrow N_2O(g) + 3H_2O(g)$$

$$\Delta H^\circ = \Delta H_f^\circ[N_2O(g)] + 3\,\Delta H_f^\circ[H_2O(g)] - 2\,\Delta H_f^\circ[NH_3(g)] - 2\,\Delta H_f^\circ[O_2(g)]$$

$$= \left(\frac{82.1\text{ kJ}}{\text{mol}} \times 1\text{ mol}\right) + \left(-\frac{241.8\text{ kJ}}{\text{mol}} \times 3\text{ mol}\right) - \left(-\frac{46.1\text{ kJ}}{\text{mol}} \times 2\text{ mol}\right) - \left(\frac{0\text{ kJ}}{\text{mol}} \times 2\text{ mol}\right)$$

$$= -551.1\text{ kJ}$$

S13-15. Write a balanced equation for the combustion reaction,

$$C_8H_{18}(\ell) + \tfrac{25}{2}O_2(g) \rightarrow 8CO_2(g) + 9H_2O(\ell)$$

and apply Hess's law:

$$\Delta H^\circ = 8\,\Delta H_f^\circ[CO_2(g)] + 9\,\Delta H_f^\circ[H_2O(\ell)] - \Delta H_f^\circ[C_8H_{18}(\ell)] - \tfrac{25}{2}\,\Delta H_f^\circ[O_2(g)]$$

$$= \left(-\frac{393.5\text{ kJ}}{\text{mol}} \times 8\text{ mol}\right) + \left(-\frac{285.8\text{ kJ}}{\text{mol}} \times 9\text{ mol}\right)$$

$$-\left(-\frac{249.9\text{ kJ}}{\text{mol}} \times 1\text{ mol}\right) - \left(\frac{0\text{ kJ}}{\text{mol}} \times 12.5\text{ mol}\right)$$

$$= -5470.3\text{ kJ}$$

All substances are chosen to be in their standard states at 25°C. The equivalent calculation for

$$C_8H_{18}(\ell) + \tfrac{25}{2}O_2(g) \rightarrow 8CO_2(g) + 9H_2O(g)$$

yields $\Delta H^\circ = -5074.3$ kJ.

S13-16. First, calculate the molar change in enthalpy:

$$Ag(s) \rightarrow Ag(g)$$

$$\Delta H^\circ = \Delta H_f^\circ[Ag(g)] - \Delta H_f^\circ[Ag(s)] = (284.9 - 0) \text{ kJ mol}^{-1} = 284.9 \text{ kJ mol}^{-1}$$

Second, calculate the heat required for 100 g:

$$100.0 \text{ g Ag} \times \frac{1 \text{ mol Ag}}{107.8682 \text{ g Ag}} \times \frac{284.9 \text{ kJ}}{\text{mol Ag}} = 264.1 \text{ kJ}$$

S13-17. Here we consider the condensation of benzene:

$$C_6H_6(g) \rightarrow C_6H_6(\ell)$$

(a) The molar change in enthalpy is obtained by a Hess's law summation:

$$\Delta H^\circ = \Delta H_f^\circ[C_6H_6(\ell)] - \Delta H_f^\circ[C_6H_6(g)] = (49.0 - 82.9) \text{ kJ mol}^{-1} = -33.9 \text{ kJ mol}^{-1}$$

After that, we calculate the heat released for a specific mass:

$$456 \text{ g } C_6H_6 \times \frac{1 \text{ mol } C_6H_6}{78.114 \text{ g } C_6H_6} \times \left(-\frac{33.9 \text{ kJ}}{\text{mol } C_6H_6}\right) = -198 \text{ kJ}$$

(b) Condensing under constant pressure, the benzene system transfers a certain quantity of heat to a certain volume of water (*n* moles):

$$q_P = nc_P\,\Delta T \equiv -\Delta H^\circ$$

$$\Delta T = \frac{-\Delta H^\circ}{nc_P} = \frac{198 \text{ kJ}}{\left(5.00 \times 10^3 \text{ mL H}_2\text{O} \times \frac{0.997 \text{ g H}_2\text{O}}{\text{mL H}_2\text{O}} \times \frac{1 \text{ mol}}{18.015 \text{ g H}_2\text{O}}\right)\left(\frac{75.3 \text{ J}}{\text{mol K}} \times \frac{1 \text{ kJ}}{1000 \text{ J}}\right)}$$

$$= 9.50 \text{ K}$$

The change in temperature is positive (and $q_P = -\Delta H^\circ$), since heat flows *out* from the reacting system to the water:

$$T_2 = T_1 + \Delta T = 25.00°\text{C} + 9.50°\text{C} = 34.50°\text{C}$$

Values for the density and heat capacity of water are available in Table C-15 of *Principles of Chemistry*.

S13-18. Use the same method as in the preceding exercise.

(a) First, calculate the enthalpy of vaporization per mole of ethanol:

$$C_2H_5OH(\ell) \rightarrow C_2H_5OH(g)$$

$$\begin{aligned}\Delta H^\circ &= \Delta H_f^\circ[C_2H_5OH(g)] - \Delta H_f^\circ[C_2H_5OH(\ell)] \\ &= [-235.1 - (-277.7)]\text{ kJ mol}^{-1} \\ &= 42.6\text{ kJ mol}^{-1}\end{aligned}$$

Second, calculate the heat required for a specific volume:

$$755\text{ mL } C_2H_5OH \times \frac{0.7873\text{ g } C_2H_5OH}{\text{mL } C_2H_5OH} \times \frac{1\text{ mol } C_2H_5OH}{46.069\text{ g } C_2H_5OH} \times \frac{42.6\text{ kJ}}{\text{mol } C_2H_5OH} = 550\text{ kJ}$$

(b) We identify q_P as $-\Delta H^\circ$, choosing the negative sign to indicate a helium-based system:

$$\Delta T = \frac{q_P}{nc_P} = \frac{-550\text{ kJ}}{(1000\text{ mol})\left(\dfrac{20.8\text{ J}}{\text{mol K}} \times \dfrac{1\text{ kJ}}{1000\text{ J}}\right)} = -26.4\text{ K}$$

The temperature decreases by 26.4 degrees, from 25.0°C to −1.4°C.

See Table C-16 for the molar heat capacity of helium gas.

S13-19. Since all we do is calculate *differences* in enthalpy, a systematic shift in each of the tabulated values would have no effect on any computation. There is no scale of absolute enthalpy.

S13-20. Heat, in general, is not a function of state. Instead, only the sum of heat and work together (the internal energy) is independent of path:

$$\Delta E = q + w$$

Under certain circumstances, however, either heat or work can indeed behave as a state function. For example, according to the definition of enthalpy we have

$$\Delta H = \Delta E + P\,\Delta V = q + w + P\,\Delta V = q - P\,\Delta V + P\,\Delta V = q$$

at constant pressure, and thus we see: The change in enthalpy, a state function, is equal to the heat exchanged during an isobaric process. It is a special case.

S13-21. An endothermic reaction draws heat away from the surroundings (the beaker) and into the system (the reacting species in solution). The beaker grows cold to the touch as heat from the outside flows into the solution.

S13-22. The equilibrium constant for an exothermic reaction decreases with increasing temperature, since heat plays the role of a product:

$$\mathrm{R} \rightarrow \mathrm{P} + \text{heat}$$

To add heat by raising the temperature is to impose a stress on the equilibrium, which is then relieved by a shift to the left—toward reactants. *Less* material is expected to dissolve at the higher temperature.

S13-23. Apply Hess's law first to the reaction as written,

$$a\mathrm{A} + b\mathrm{B} \rightarrow c\mathrm{C} + d\mathrm{D}$$

$$\Delta H^\circ(\text{forward}) = c\,\Delta H_f^\circ(\mathrm{C}) + d\,\Delta H_f^\circ(\mathrm{D}) - a\,\Delta H_f^\circ(\mathrm{A}) - b\,\Delta H_f^\circ(\mathrm{B})$$

and then to the process taken in the opposite direction:

$$c\mathrm{C} + d\mathrm{D} \rightarrow a\mathrm{A} + b\mathrm{B}$$

$$\Delta H^\circ(\text{reverse}) = a\,\Delta H_f^\circ(\mathrm{A}) + b\,\Delta H_f^\circ(\mathrm{B}) - c\,\Delta H_f^\circ(\mathrm{C}) - d\,\Delta H_f^\circ(\mathrm{D}) = -\Delta H^\circ(\text{forward})$$

S13-24. Sublimation refers to the conversion of a substance from solid into gas:

$$\mathrm{I_2(s)} \rightarrow \mathrm{I_2(g)}$$

$$\Delta H^\circ = \Delta H_f^\circ[\mathrm{I_2(g)}] - \Delta H_f^\circ[\mathrm{I_2(s)}] = (62.4 - 0)\ \text{kJ mol}^{-1} = 62.4\ \text{kJ mol}^{-1}$$

S13-25. By definition, an element in its standard state is assigned the value $\Delta H_f^\circ = 0$. Molecular hydrogen therefore has a standard enthalpy of formation equal to zero, the difference in enthalpy between the product (gaseous H_2) and its component elements in their normal states (also gaseous H_2). The defining calculation is tautological:

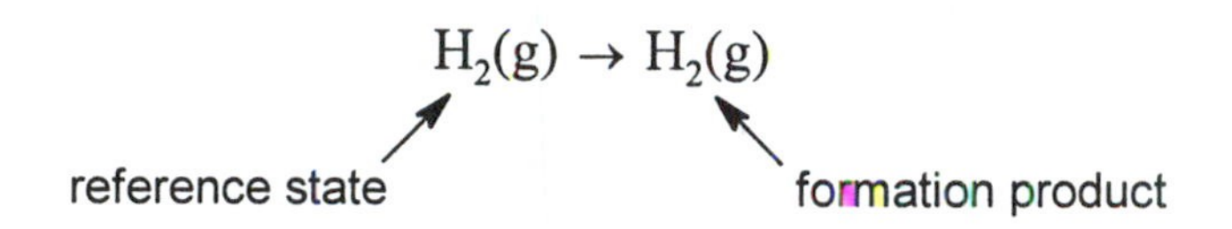

$$\Delta H^\circ = \Delta H_f^\circ[H_2(g)] - \Delta H_f^\circ[H_2(g)] = \left(\frac{0\ \text{kJ}}{\text{mol}} \times 1\ \text{mol}\right) - \left(\frac{0\ \text{kJ}}{\text{mol}} \times 1\ \text{mol}\right) = 0$$

Note that the reactant in the process

$$2H(g) \rightarrow H_2(g) \qquad \Delta H^\circ = -436\ \text{kJ}$$

is not molecular hydrogen (the standard state) but rather *atomic* hydrogen. The hydrogen atom contributes a nonzero enthalpy of formation to the reaction:

$$\Delta H^\circ = \Delta H_f^\circ[H_2(g)] - 2\,\Delta H_f^\circ[H(g)] = \left(\frac{0\ \text{kJ}}{\text{mol}} \times 1\ \text{mol}\right) - \left(\frac{218\ \text{kJ}}{\text{mol}} \times 2\ \text{mol}\right) = -436\ \text{kJ}$$

Chapter 14

Free Energy and the Direction of Change

S14-1. Suppose that a coin has three sides: heads (H), tails (T), and feet (F). **(a)** How many outcomes are possible when the coin is tossed twice? List them. **(b)** Which outcome or outcomes are more probable than the rest?

S14-2. System 1 absorbs 100 joules of heat reversibly at a temperature of 5 K. System 2 also absorbs 100 joules reversibly, but at a temperature of 500 K. **(a)** For each system, state whether the change in entropy is positive, negative, or zero. **(b)** Which system realizes the greater change in entropy?

S14-3. Say that a system of gas exchanges thermal energy reversibly with a heat reservoir at 100 K. **(a)** If the entropy of the system decreases by 10 J K^{-1}, what quantity of heat is exchanged? **(b)** In which direction does the heat flow—from the reservoir to the gas, or from the gas to the reservoir?

S14-4. One mole of gas absorbs 100 joules of heat reversibly from a reservoir at 25°C. Calculate the change in entropy for the gas.

S14-5. A frictionless system, thermally isolated from its surroundings, is subjected to a reversible mechanical process. The work done is equal to 1 kJ. **(a)** Calculate ΔE.
(b) Calculate ΔS.

S14-6. A system undergoes a series of reversible changes in which heat flows neither in nor out. **(a)** Does the entropy of the system increase, decrease, or remain the same?
(b) Does the entropy of the surroundings increase, decrease, or remain the same?
(c) Does the entropy of the universe increase, decrease, or remain the same?

S14-7. A system spontaneously undergoes an irreversible endothermic reaction. **(a)** Are you able to determine whether the change in entropy for the system is positive, negative or zero? **(b)** If yes, write a symbolic expression for ΔS in terms of other thermodynamic variables. If no, explain why not.

S14-8. As above, a system spontaneously undergoes an irreversible endothermic reaction. **(a)** Does the entropy of the surroundings increase, decrease, or remain the same? **(b)** Does the entropy of the universe increase, decrease, or remain the same?

S14-9. A system spontaneously undergoes an irreversible exothermic reaction. **(a)** Does the entropy of the surroundings increase, decrease, or remain the same? **(b)** Does the entropy of the universe increase, decrease, or remain the same?

S14-10. A confined gas undergoes a reversible change in internal energy, ΔE_{rev}, at constant volume. What is the corresponding change in entropy for the surroundings? Assume that system and surroundings are in thermal equilibrium.

S14-11. If pressure and temperature are constant, then the quantity $-\Delta G/T$ registers the change in entropy for system and surroundings combined—the universe. Recall, in particular, that a change in the Gibbs free energy under these conditions is defined as

$$\Delta G = \Delta H - T\,\Delta S$$

Now consider a change in the *Helmholtz* free energy, stated provisionally as

$$\Delta A = \Delta E - T\,\Delta S$$

Show that the quantity $-\Delta A/T$ registers the global change in entropy for a transformation undergone by a system at constant volume and temperature. Assume that system and surroundings are in thermal equilibrium.

S14-12. Assume that $\Delta H^\circ > 0$ and $K > 1$ for the hypothetical process

$$\text{A} + \text{B} \rightleftarrows \text{C} + \text{D}$$

at a certain temperature T. What is the algebraic sign of ΔS°?

S14-13. Assume that $\Delta H^\circ < 0$ and $K < 1$ for the hypothetical process

$$\text{A} \rightleftarrows \text{B}$$

at a certain temperature T. Which species has the higher standard entropy, A or B?

S14-14. Assume that $\Delta H^\circ = 100$ kJ and $\Delta S^\circ = 100$ J K^{-1} for the hypothetical process

$$\text{A} + 2\text{B} + 5\text{C} \rightleftarrows 3\text{D} + 4\text{E}$$

(a) Calculate the equilibrium constant at 1000 K. **(b)** Over what range of temperature, if any, is the reaction spontaneous?

S14-15. Which species in each pair has the higher value of ΔG_f° at 298 K?

(a) $H_2(g)$, $H_2(\ell)$ **(b)** $H_2O(g)$, $H_2O(\ell)$ **(c)** $Na(\ell)$, $Na(s)$

Try to answer without consulting Appendix C.

S14-16. Again, but consider a higher temperature: Which species has the greater value of ΔG_f° at 450 K?

(a) $H_2(g)$, $H_2(\ell)$ **(b)** $H_2O(g)$, $H_2O(\ell)$ **(c)** $Na(\ell)$, $Na(s)$

Try to answer without consulting Appendix C. Note from Table C-9, however, that the melting point of sodium is 97.7°C.

S14-17. (a) Calculate ΔH°, ΔS°, and ΔG° for the reaction below:

$$C_6H_{12}O_6(s) + 6O_2(g) \rightleftarrows 6CO_2(g) + 6H_2O(\ell)$$

(b) Over what range of temperature is the process spontaneous?

S14-18. Calculate ΔG_f° for $N_2O_5(g)$ given the information below:

$$N_2O_5(g) + H_2O(\ell) \rightleftarrows 2HNO_3(\ell) \qquad \Delta G^\circ = -39.3 \text{ kJ}$$

Other relevant data may be found in Appendix C.

S14-19. Calculate ΔH°, ΔS°, ΔG°, and K for the reaction below:

$$N_2(g) + 3O_2(g) + H_2(g) \rightleftarrows 2HNO_3(\ell)$$

Assume a temperature of 298 K.

S14-20. Use the data in Appendix C to estimate the boiling temperature of carbon tetrachloride, CCl_4. (The published boiling point of CCl_4 is 76.8°C, or 350 K.)

S14-21. The normal boiling point of methanol (CH_3OH) is 64.6°C, and the standard entropy of vaporization is 112.9 J mol^{-1} K^{-1}. **(a)** Without consulting Appendix C, estimate ΔH° for the reaction below:

$$CH_3OH(g) \rightleftarrows CH_3OH(\ell)$$

(b) Calculate ΔH° directly from the appropriate data in Appendix C. Compare the two results.

S14-22. Thermodynamic data for substances A, B, C, and D are tabulated below:

	ΔG_f° (kJ mol^{-1})
A	50
B	100
C	200
D	150

(a) Calculate the value of ΔG (not ΔG°) when the reaction

$$A + 3B \rightleftarrows 2C + D$$

comes to equilibrium at 25°C. **(b)** Is the equilibrium constant greater than 1, less than 1, or equal to 1 at 25°C?

S14-23. Use the hypothetical formation data from the preceding exercise to solve the following problem: **(a)** Calculate the value of ΔG when the reaction

$$4A + 6B \rightleftarrows C + D$$

comes to equilibrium at 25°C. **(b)** Is the equilibrium constant greater than 1, less than 1, or equal to 1 at 25°C? Compare the results with those of the preceding exercise.

SOLUTIONS

S14-1. We treat the case of an unloaded three-sided "coin." Each of the three sides is equally likely to turn up on a given throw.

(a) There are nine possible outcomes:

HH	TH	FH
HT	TT	FT
HF	TF	FF

(b) The outcomes may be grouped into six distributions:

I	HT	TH	one head, one tail
II	HF	FH	one head, one foot
III	TF	FT	one tail, one foot
IV	HH		two heads
V	TT		two tails
VI	FF		two feet

Each of the distributions I, II, and III (probability = 2/9) is twice as likely as each of the distributions IV, V, and VI (probability = 1/9).

S14-2. We consider a reversible flow of heat at constant temperature, an idealized process. The corresponding change in entropy depends on both the quantity of heat and the absolute temperature:

$$\Delta S = \frac{q_{\text{rev}}}{T}$$

(a) The change in entropy is positive for each system. Heat is absorbed:

$$\Delta S_1 = \frac{100 \text{ J}}{5 \text{ K}} = 20 \text{ J K}^{-1}$$

$$\Delta S_2 = \frac{100 \text{ J}}{500 \text{ K}} = 0.2 \text{ J K}^{-1}$$

(b) System 1 realizes the greater change in entropy. The same amount of heat flows into each system, but system 1—at a lower temperature—is more ordered initially and hence more susceptible to disruption.

S14-3. Another application of the defining equation for entropy:

$$\Delta S = \frac{q_{\text{rev}}}{T}$$

(a) Given ΔS and T, we solve for q_{rev}:

$$q_{\text{rev}} = T\,\Delta S = (100 \text{ K})(-10 \text{ J K}^{-1}) = -1000 \text{ J}$$

(b) Heat flows reversibly from the gas to the reservoir, and the system becomes more ordered as a result. The change in entropy is negative.

S14-4. Remember to use the absolute—Kelvin—temperature in all thermodynamic calculations, not the Celsius temperature:

$$\Delta S = \frac{q_{\text{rev}}}{T} = \frac{100 \text{ J}}{298 \text{ K}} = 0.336 \text{ J K}^{-1}$$

S14-5. The heat transferred is zero ($q_{\text{rev}} = 0$), since the system is thermally isolated.

(a) $\Delta E = q + w = w = 1$ kJ

(b) $\Delta S = \frac{q_{\text{rev}}}{T} = 0$

S14-6. The problem describes an idealized process in which all changes are reversible and no microscopic disordering takes place. There is no passage of heat between system and surroundings.

(a) $\Delta S_{\text{sys}} = \frac{q_{\text{rev, sys}}}{T} = 0$

(b) $\Delta S_{\text{surr}} = \frac{q_{\text{rev, surr}}}{T} = 0$

(c) $\Delta S_{\text{univ}} = \Delta S_{\text{sys}} + \Delta S_{\text{surr}} = 0$

S14-7. The Gibbs free energy decreases during a spontaneous reaction:

$$\Delta G = \Delta H - T\Delta S < 0 \qquad \text{(spontaneous)}$$

(a) If ΔH for the system is positive (as in an endothermic process), then ΔS for the system must be positive in order for ΔG to be negative: $\Delta S > 0$.

(b) Rearrange the defining equation for ΔG:

$$\Delta S = \frac{\Delta H - \Delta G}{T}$$

S14-8. Dividing ΔG by $-T$,

$$-\frac{\Delta G}{T} = -\frac{\Delta H_{\text{sys}}}{T} + \Delta S_{\text{sys}}$$

we identify the term $-\Delta H_{\text{sys}}/T$ as the entropy change for the surroundings:

$$\Delta S_{\text{surr}} = -\frac{\Delta H_{\text{sys}}}{T}$$

The quantity $-\Delta H_{\text{sys}}$ (opposite in sign to ΔH_{sys}) represents a reversible flow of heat as viewed from the surroundings. Together with the entropy change for the system alone, we then have the global change ΔS_{univ}:

$$\Delta S_{\text{univ}} = -\frac{\Delta G}{T} = \Delta S_{\text{surr}} + \Delta S_{\text{sys}}$$

(a) $\Delta S_{\text{surr}} = -\frac{\Delta H_{\text{sys}}}{T} < 0 \quad \text{if} \quad \Delta H_{\text{sys}} > 0$

(b) $\Delta S_{\text{univ}} > 0$ (true for any spontaneous process)

S14-9. We apply the same reasoning as in the previous exercise, this time for a spontaneous exothermic reaction.

(a) $\Delta S_{\text{surr}} = -\frac{\Delta H_{\text{sys}}}{T} > 0 \quad \text{if} \quad \Delta H_{\text{sys}} < 0$

(b) $\Delta S_{\text{univ}} > 0$ (true for any spontaneous process)

S14-10. The work done is zero since the process takes place at constant volume. As a result, the heat transferred must be equal to the change in internal energy:

$$\Delta E_{\text{rev}} = q_{\text{rev}} + w = q_{\text{rev}}$$

$$\Delta S_{\text{sys}} = \frac{q_{\text{rev}}}{T} = \frac{\Delta E_{\text{rev}}}{T}$$

We assume, because system and surroundings are in thermal equilibrium, that the reversible flow of heat to the surroundings is equal and opposite:

$$\Delta S_{\text{surr}} = -\frac{q_{\text{rev}}}{T} = -\frac{\Delta E_{\text{rev}}}{T}$$

S14-11. For clarity, use the subscript *sys* explicitly to label the thermodynamic functions A, E, and S:

$$\Delta A_{\text{sys}} = \Delta E_{\text{sys}} - T\Delta S_{\text{sys}}$$

$$-\frac{\Delta A_{\text{sys}}}{T} = -\frac{\Delta E_{\text{sys}}}{T} + \Delta S_{\text{sys}}$$

Since the process occurs at constant volume, the change in internal energy is due entirely to a flow of heat:

$$\Delta E_{sys} = q + w = q$$

With system and surroundings in thermal equilibrium we then interpret $-\Delta E_{sys}$ as a reversible flow of heat into the surroundings. Thus the associated change in entropy for the surroundings is

$$\Delta S_{surr} = -\frac{\Delta E_{sys}}{T}$$

and the global change in entropy is

$$-\frac{\Delta A_{sys}}{T} = \Delta S_{surr} + \Delta S_{sys}$$

S14-12. Since $K > 1$, we know that ΔG° is negative:

$$\Delta G^\circ = -RT \ln K = \Delta H^\circ - T\Delta S^\circ < 0 \qquad \text{if} \qquad K > 1$$

If ΔH° is positive, then ΔS° must be positive as well.

S14-13. Similar. With $K < 1$, the sign of ΔG° is now positive:

$$\Delta G^\circ = -RT \ln K = \Delta H^\circ - T\Delta S^\circ > 0 \qquad \text{if} \qquad K < 1$$

Given that ΔH° is negative, we infer that ΔS° must also be negative if ΔG° is to be positive. The reactant A therefore has a higher standard entropy than the product B:

$$\Delta S^\circ = S^\circ(\text{B}) - S^\circ(\text{A}) < 0$$

$$S^\circ(\text{A}) > S^\circ(\text{B})$$

S14-14. Once again, we turn to the master equation for the Gibbs free energy:

$$\Delta G^\circ = \Delta H^\circ - T\Delta S^\circ = -RT \ln K$$

(a) The equilibrium constant

$$K = \exp\left(-\frac{\Delta G^\circ}{RT}\right)$$

is related exponentially to the standard change in free energy. With $\Delta G^\circ = 0$ at 1000 K,

$$\Delta G^\circ = \Delta H^\circ - T\Delta S^\circ = \left(100 \text{ kJ} \times \frac{1000 \text{ J}}{\text{kJ}}\right) - (1000 \text{ K})(100 \text{ J K}^{-1}) = 0$$

the equilibrium constant has unit value:

$$K = \exp\left(-\frac{\Delta G^\circ}{RT}\right) = \exp 0 = 1 \qquad (T = 1000 \text{ K})$$

(b) ΔG° turns negative—and K becomes greater than 1—at all temperatures above 1000 K:

$$\Delta G^\circ = \Delta H^\circ - T\Delta S^\circ < 0 \qquad (\Delta H^\circ > 0,\ \Delta S^\circ > 0,\ T > T_0)$$

$$T_0 = \frac{\Delta H^\circ}{\Delta S^\circ} = \frac{100 \text{ kJ} \times \dfrac{1000 \text{ J}}{\text{kJ}}}{100 \text{ J K}^{-1}} = 1000 \text{ K}$$

S14-15. Spontaneity lies in the direction that yields the more stable species—the one with the lower free energy of formation.

(a) Molecular hydrogen, normally a gas at 298 K, is produced spontaneously by vaporization:

$$H_2(\ell) \rightleftarrows H_2(g)$$

The liquid form of hydrogen, $H_2(\ell)$, has the higher free energy.

(b) Water condenses spontaneously at 298 K. The vapor lies higher in free energy than the liquid, and thus the process

$$H_2O(g) \rightleftarrows H_2O(\ell)$$

is thermodynamically favored under the stated conditions.

(c) Since sodium exists naturally as a solid at 298 K, the freezing transition proceeds spontaneously:

$$Na(\ell) \rightleftarrows Na(s)$$

The liquid form, $Na(\ell)$, has the higher free energy.

S14-16. Determine, as in the preceding exercise, which of the two states exists naturally at the given temperature.

(a) $H_2(g)$ is the normal state of hydrogen at 450 K. The liquid form, $H_2(\ell)$, therefore has the higher free energy.

(b) Water exists normally as a vapor at 450 K (177°C). The liquid form, $H_2O(\ell)$, has the higher free energy.

(c) Since sodium melts at 371 K, the liquid state is favored at 450 K. Metallic sodium, Na(s), has the higher free energy.

S14-17. We consider the combustion of glucose to carbon dioxide and water:

$$C_6H_{12}O_6(s) + 6O_2(g) \rightleftarrows 6CO_2(g) + 6H_2O(\ell)$$

(a) Summation of the formation data over products and reactants yields the net change in each quantity:

$$\Delta H^\circ = 6\Delta H_f^\circ[CO_2(g)] + 6\Delta H_f^\circ[H_2O(\ell)] - \Delta H_f^\circ[C_6H_{12}O_6(s)] - 6\Delta H_f^\circ[O_2(g)]$$

$$= \left(-\frac{393.5 \text{ kJ}}{\text{mol}} \times 6 \text{ mol}\right) + \left(-\frac{285.8 \text{ kJ}}{\text{mol}} \times 6 \text{ mol}\right)$$

$$-\left(-\frac{1274.4 \text{ kJ}}{\text{mol}} \times 1 \text{ mol}\right) - \left(\frac{0 \text{ kJ}}{\text{mol}} \times 6 \text{ mol}\right)$$

$$= -2801.4 \text{ kJ}$$

$$\Delta S^\circ = 6S^\circ[CO_2(g)] + 6S^\circ[H_2O(\ell)] - S^\circ[C_6H_{12}O_6(s)] - 6S^\circ[O_2(g)]$$

$$= \left(\frac{213.6 \text{ J}}{\text{mol K}} \times 6 \text{ mol}\right) + \left(\frac{70.0 \text{ J}}{\text{mol K}} \times 6 \text{ mol}\right) - \left(\frac{212.1 \text{ J}}{\text{mol K}} \times 1 \text{ mol}\right) - \left(\frac{205.0 \text{ J}}{\text{mol K}} \times 6 \text{ mol}\right)$$

$$= 259.5 \text{ J K}^{-1}$$

$$\Delta G^\circ = 6\Delta G_f^\circ[CO_2(g)] + 6\Delta G_f^\circ[H_2O(\ell)] - \Delta G_f^\circ[C_6H_{12}O_6(s)] - 6\Delta G_f^\circ[O_2(g)]$$

$$= \left(-\frac{394.4\text{ kJ}}{\text{mol}} \times 6\text{ mol}\right) + \left(-\frac{237.2\text{ kJ}}{\text{mol}} \times 6\text{ mol}\right)$$

$$-\left(-\frac{910.1\text{ kJ}}{\text{mol}} \times 1\text{ mol}\right) - \left(\frac{0\text{ kJ}}{\text{mol}} \times 6\text{ mol}\right)$$

$$= -2879.5\text{ kJ}$$

(b) The reaction is spontaneous at all temperatures, since $\Delta H^\circ < 0$ and $\Delta S^\circ > 0$.

S14-18. Dinitrogen pentoxide, N_2O_5, is not one of the compounds listed in Table C-16 of *Principles of Chemistry*. To determine its standard free energy of formation, we use data known for the reaction

$$N_2O_5(g) + H_2O(\ell) \rightleftarrows 2HNO_3(\ell) \qquad \Delta G^\circ = -39.3\text{ kJ}$$

and work backward:

$$\Delta G^\circ = 2\Delta G_f^\circ[HNO_3(\ell)] - \Delta G_f^\circ[N_2O_5(g)] - \Delta G_f^\circ[H_2O(\ell)]$$

$$\Delta G_f^\circ[N_2O_5(g)] = 2\Delta G_f^\circ[HNO_3(\ell)] - \Delta G_f^\circ[H_2O(\ell)] - \Delta G^\circ$$

$$= [2(-80.7) - (-237.2) - (-39.3)]\text{ kJ mol}^{-1}$$

$$= 115.1\text{ kJ mol}^{-1}$$

S14-19. Here we analyze the formation of nitric acid from its elements:

$$N_2(g) + 3O_2(g) + H_2(g) \rightleftarrows 2HNO_3(\ell)$$

The enthalpy of the system goes down:

$$\Delta H^\circ = 2\Delta H_f^\circ[HNO_3(\ell)] - \Delta H_f^\circ[N_2(g)] - 3\Delta H_f^\circ[O_2(g)] - \Delta H_f^\circ[H_2(g)]$$

$$= \left(-\frac{174.1\text{ kJ}}{\text{mol}} \times 2\text{ mol}\right) - \left(\frac{0\text{ kJ}}{\text{mol}} \times 1\text{ mol}\right) - \left(\frac{0\text{ kJ}}{\text{mol}} \times 3\text{ mol}\right) - \left(\frac{0\text{ kJ}}{\text{mol}} \times 1\text{ mol}\right)$$

$$= -348.2\text{ kJ}$$

The entropy of the system goes down as well:

$$\Delta S^\circ = 2S^\circ[\text{HNO}_3(\ell)] - S^\circ[\text{N}_2(\text{g})] - 3S^\circ[\text{O}_2(\text{g})] - S^\circ[\text{H}_2(\text{g})]$$

$$= \left(\frac{155.6 \text{ J}}{\text{mol K}} \times 2 \text{ mol}\right) - \left(\frac{191.5 \text{ J}}{\text{mol K}} \times 1 \text{ mol}\right) - \left(\frac{205.0 \text{ J}}{\text{mol K}} \times 3 \text{ mol}\right) - \left(\frac{130.6 \text{ J}}{\text{mol K}} \times 1 \text{ mol}\right)$$

$$= -625.9 \text{ J K}^{-1}$$

The standard change in free energy is strongly negative at 298 K, and the reaction is spontaneous:

$$\Delta G^\circ = 2\Delta G_f^\circ[\text{HNO}_3(\ell)] - \Delta G_f^\circ[\text{N}_2(\text{g})] - 3\Delta G_f^\circ[\text{O}_2(\text{g})] - \Delta G_f^\circ[\text{H}_2(\text{g})]$$

$$= \left(-\frac{80.7 \text{ kJ}}{\text{mol}} \times 2 \text{ mol}\right) - \left(\frac{0 \text{ kJ}}{\text{mol}} \times 1 \text{ mol}\right) - \left(\frac{0 \text{ kJ}}{\text{mol}} \times 3 \text{ mol}\right) - \left(\frac{0 \text{ kJ}}{\text{mol}} \times 1 \text{ mol}\right)$$

$$= -161.4 \text{ kJ} \quad (\text{per mole of N}_2)$$

The equilibrium constant is considerably greater than 1:

$$K = \exp\left(-\frac{\Delta G^\circ}{RT}\right) = \exp\left[-\frac{\left(-161.4 \text{ kJ mol}^{-1} \times \frac{1000 \text{ J}}{\text{kJ}}\right)}{(8.3145 \text{ J mol}^{-1} \text{ K}^{-1})(298 \text{ K})}\right] = 1.95 \times 10^{28}$$

S14-20. The change in free energy for the vaporization reaction

$$\text{CCl}_4(\ell) \rightleftarrows \text{CCl}_4(\text{g})$$

is zero at the boiling temperature, T_b:

$$\Delta G_{\text{vap}} = \Delta H_{\text{vap}}^\circ - T_b \, \Delta S_{\text{vap}}^\circ = 0$$

Thus we estimate the boiling temperature (using, for simplicity, standard values at 25°C):

$$T_b = \frac{\Delta H_{\text{vap}}^\circ}{\Delta S_{\text{vap}}^\circ} = \frac{\Delta H_f^\circ[\text{CCl}_4(\text{g})] - \Delta H_f^\circ[\text{CCl}_4(\ell)]}{S^\circ[\text{CCl}_4(\text{g})] - S^\circ[\text{CCl}_4(\ell)]} = \frac{[-102.9 - (-135.4)] \text{ kJ}}{(0.3097 - 0.2164) \text{ kJ K}^{-1}} = 348 \text{ K}$$

S14-21. Similar. Here we consider the condensation–vaporization equilibrium of methanol ($T_b = 64.6°C = 337.8$ K):

$$CH_3OH(g) \rightleftarrows CH_3OH(\ell) \qquad \Delta S^\circ_{con} = -\Delta S^\circ_{vap} = -112.9 \text{ J mol}^{-1} \text{ K}^{-1}$$

(a) At equilibrium, the free energies of the two phases are equal:

$$\Delta G_{con} = \Delta H^\circ_{con} - T_b \, \Delta S^\circ_{con} = 0$$

$$\Delta H^\circ_{con} = T_b \, \Delta S^\circ_{con} = (337.8 \text{ K})(-0.1129 \text{ kJ mol}^{-1} \text{ K}^{-1}) = -38.1 \text{ kJ mol}^{-1}$$

(b) Direct application of Hess's law yields effectively the same result:

$$\begin{aligned} \Delta H^\circ_{con} &= \Delta H^\circ_f[CH_3OH(\ell)] - \Delta H^\circ_f[CH_3OH(g)] \\ &= [-238.7 - (-200.7)] \text{ kJ mol}^{-1} \\ &= -38.0 \text{ kJ mol}^{-1} \end{aligned}$$

S14-22. We have a hypothetical reaction involving four substances:

$$A + 3B \rightleftarrows 2C + D$$

(a) At equilibrium, there is no drive to produce one species at the expense of another. The difference in free energy between products and reactants is zero: $\Delta G = 0$.

(b) We sum over the standard free energies of formation, as given, to obtain the free energy of reaction at 25°C:

$$\begin{aligned} \Delta G^\circ &= 2\Delta G^\circ_f(C) + \Delta G^\circ_f(D) - \Delta G^\circ_f(A) - 3\Delta G^\circ_f(B) \\ &= \left(\frac{200 \text{ kJ}}{\text{mol}} \times 2 \text{ mol}\right) + \left(\frac{150 \text{ kJ}}{\text{mol}} \times 1 \text{ mol}\right) - \left(\frac{50 \text{ kJ}}{\text{mol}} \times 1 \text{ mol}\right) - \left(\frac{100 \text{ kJ}}{\text{mol}} \times 3 \text{ mol}\right) \\ &= 200 \text{ kJ} \end{aligned}$$

The equilibrium constant at this temperature is less than 1, since ΔG° is positive:

$$K = \exp\left(-\frac{\Delta G^\circ}{RT}\right) < 1 \qquad \text{if} \qquad \Delta G^\circ > 0$$

S14-23. Carry over the hypothetical formation data used in Exercise S14-22, this time for a different reaction:

$$4\text{A} + 6\text{B} \rightleftarrows \text{C} + \text{D}$$

(a) Again: ΔG, the instantaneous difference in free energy, is zero for any chemical system in equilibrium.

(b) The *standard* difference in free energy, referring to species at 1 atm or 1 *M*, is different for every reaction. Here the value of ΔG° is negative,

$$\begin{aligned}\Delta G^\circ &= \Delta G_\text{f}^\circ(\text{C}) + \Delta G_\text{f}^\circ(\text{D}) - 4\,\Delta G_\text{f}^\circ(\text{A}) - 6\,\Delta G_\text{f}^\circ(\text{B}) \\ &= \left(\frac{200\text{ kJ}}{\text{mol}} \times 1\text{ mol}\right) + \left(\frac{150\text{ kJ}}{\text{mol}} \times 1\text{ mol}\right) - \left(\frac{50\text{ kJ}}{\text{mol}} \times 4\text{ mol}\right) - \left(\frac{100\text{ kJ}}{\text{mol}} \times 6\text{ mol}\right) \\ &= -450\text{ kJ}\end{aligned}$$

and the equilibrium constant at 298 K is greater than 1:

$$K = \exp\left(-\frac{\Delta G^\circ}{RT}\right) > 1 \qquad \text{if} \qquad \Delta G^\circ < 0$$

Chapter 15

Making Accommodations—Solubility and Molecular Recognition

S15-1. Explain the difference between the *solubility* and the *solubility-product constant* of a substance.

S15-2. One mole of X requires 30 seconds to dissolve in 1 liter of water at 10°C. One mole of Y requires 300 seconds to dissolve in the same amount of water at the same temperature. At 90°C, however, the times are equal: one mole of X dissolves in 3 seconds, and 1 mole of Y dissolves in 3 seconds as well. Note further that no undissolved solute is evident in any of the solutions considered. **(a)** Can you say whether the reactions

$$X(s) \rightleftarrows X(aq) \qquad \text{and} \qquad Y(s) \rightleftarrows Y(aq)$$

are exothermic or endothermic? If yes, do so. If no, explain why not. **(b)** Can you determine whether the solubility of X is greater than, less than, or equal to the solubility of Y at each temperature? If yes, do so. If no, explain why not. **(c)** Can you calculate K_{sp} for X and Y at each temperature? If yes, do so. If no, explain why not.

S15-3. Suppose that a saturated solution of X (volume = 1 L) is produced after 10 s of rapid stirring at 25°C. Thermodynamic data are given below:

	ΔH_f° (kJ mol^{-1})	ΔG_f° (kJ mol^{-1})
X(s)	120	100
X(aq)	140	80

(a) Can you determine whether the solubility of X at 50°C will be greater than, less than, or equal to the solubility at 25°C? If yes, do so. If no, explain why not. **(b)** Can you determine whether the saturation point at 50°C will be reached in 10 s, more than 10 s, or less than 10 s? If yes, do so. If no, explain why not.

S15-4. Consider a generic ionic compound A_jB_k, where A denotes a cation with charge p and B denotes an anion with charge $-q$. **(a)** Determine the algebraic relationship between p and q. **(b)** Write a symbolic algebraic expression for the solubility-product constant of A_jB_k. **(c)** Write an equation that gives the molar solubility of A_jB_k in terms of only j, k, and the numerical value of K_{sp}.

S15-5. Assume that the numerical value of K_{sp} at some particular temperature is the same for compounds AB_2 and AB_3. **(a)** Calculate the ratio of solubilities. **(b)** Is the concentration of cations in a saturated solution of AB_2 necessarily equal to the concentration of cations in a saturated solution of AB_3? **(c)** For saturated solutions of both AB_2 and AB_3, express the concentration of anions in terms of K_{sp}.

S15-6. The solubility of $Mg(OH)_2$ is approximately 0.0065 g L^{-1} at 25°C. Calculate K_{sp}.

S15-7. The solubility-product constant for SrF_2 is $K_{sp}(25°C) = 4.3 \times 10^{-9}$. Calculate the solubility of SrF_2 in g L^{-1} at 25°C.

S15-8. Rank the following ionic solids in order of increasing molar solubility in water:

$$Ca(NO_3)_2,\ Hg_2I_2,\ MgF_2,\ ZnCO_3$$

Assume a temperature of 25°C. Relevant information may be found in Appendix C.

S15-9. The solubility-product constant for AgCl is $K_{sp}(25°C) = 1.8 \times 10^{-10}$. **(a)** What minimum volume of solution is required to dissolve 1.0 g AgCl in water at 25°C? **(b)** What minimum volume of solution is required to dissolve 1.0×10^2 g?

S15-10. The solubility-product constant for $PbCl_2$ is $K_{sp}(25°C) = 1.6 \times 10^{-5}$. **(a)** What minimum volume of solution is required to dissolve 1.0 g $PbCl_2$ in water at 25°C? **(b)** What minimum volume of solution is required to dissolve 1.0×10^2 g?

S15-11. Assume that x grams of $MgCO_3$ can be dissolved in 100 milliliters of pure water. Will the mass of $MgCO_3$ that can be dissolved in 100 milliliters of 1.00 M Na_2SO_4 be greater than, less than, or the same as x?

S15-12. Assume, again, that x grams of $MgCO_3$ can be dissolved in 100 milliliters of pure water. Will the mass of $MgCO_3$ that can be dissolved in 100 milliliters of 1.00 M $MgSO_4$ be greater than, less than, or the same as x?

S15-13. Imagine that a saturated solution of Ag^+(aq) and Cl^-(aq) coexists in equilibrium with an infinitely large pile of AgCl(s). The temperature is 25°C. **(a)** What happens as additional water drips into the solution? **(b)** Does the concentration of AgCl(aq) increase, decrease, or remain the same as water is added?

S15-14. Consider a variation of the preceding exercise: A saturated solution of AgCl(aq) comes to equilibrium not with an infinitely large pile of AgCl(s), but rather with just a single grain. Again, the temperature is 25°C. **(a)** Is the equilibrium concentration of AgCl(aq) in this new situation greater than, less than, or equal to its previous value? **(b)** What happens as additional water drips into the solution? **(c)** Does the concentration of AgCl(aq) increase, decrease, or remain the same as more water is added?

S15-15. Suppose that a saturated solution abruptly loses 25% of its volume owing to evaporation of the solvent. **(a)** If a precipitate does not form immediately, does a heterogeneous equilibrium exist in the solution? **(b)** In the absence of precipitation, is the numerical value of Q (the reaction quotient) greater than, less than, or equal to K_{sp}? **(c)** Does the numerical value of K_{sp} change as the volume changes? **(d)** Is the *equilibrium* concentration of solute in the smaller volume greater than, less than, or equal to its value in the larger volume?

S15-16. Consider a specific numerical example to illustrate the idea developed in the previous exercise: A saturated solution of AgOH ($K_{sp} = 1.8 \times 10^{-8}$ at 25°C) loses 25% of its volume owing to evaporation. Calculate the equilibrium concentrations of Ag^+ and OH^- before and after evaporation.

S15-17. Study the thermodynamic data given below for two hypothetical equilibria:

Reaction	$\Delta H°$ (kJ mol^{-1})	$\Delta S°$ (J mol^{-1} K^{-1})
$AB(s) \rightleftarrows A^+(aq) + B^-(aq)$	50	100
$CD(s) \rightleftarrows C^+(aq) + D^-(aq)$	−50	100

(a) Which of the two ionic solutes—AB or CD—is more soluble at 25°C? **(b)** Does the solubility of AB increase, decrease, or remain the same at temperatures greater than 25°C? **(c)** Similarly, for CD: Does the solubility increase, decrease, or remain the same at higher temperatures? Account for the direction of change.

S15-18. Another pair of hypothetical equilibria:

Reaction	$\Delta H°$ (kJ mol^{-1})	$\Delta S°$ (J mol^{-1} K^{-1})
$AB(s) \rightleftarrows A^+(aq) + B^-(aq)$	50	100
$CD(s) \rightleftarrows C^+(aq) + D^-(aq)$	−50	−300

(a) Which of the two ionic solutes—AB or CD—is more soluble at 25°C? **(b)** Does the solubility of AB increase, decrease, or remain the same at temperatures greater than 25°C? **(c)** Does the solubility of CD increase, decrease, or remain the same at higher temperatures? Account for the direction of change.

S15-19. Study the thermodynamic data given below for two hypothetical equilibria:

REACTION	$\Delta H°$ (kJ mol^{-1})	$\Delta S°$ (J mol^{-1} K^{-1})
$AB(s) \rightleftarrows A^+(aq) + B^-(aq)$	50	100
$CD(s) \rightleftarrows C^+(aq) + D^-(aq)$	5	−100

(a) Calculate the ratio $K_{sp}(AB)/K_{sp}(CD)$ at 25°C. **(b)** Calculate the ratio $K_{sp}(AB)/K_{sp}(CD)$ at 4°C. **(c)** Account for the different ratios at the different temperatures.

S15-20. Use the formation data in Appendix C to calculate K_{sp} for Li_2CO_3 at 25°C.

S15-21. Would you expect K_{sp} for $Ca(NO_3)_2$ to be greater than 1, less than 1, or equal to 1? Relevant data may be found in Appendix C.

S15-22. Two halves of a glass vessel are separated by a semipermeable membrane open only to the passage of water molecules, not sodium ions. On the left-hand side, the concentration of Na^+ has the value x. On the right-hand side, an experimenter adjusts the initial concentration of Na^+ to the value $x/2$. The volume of liquid in each half is 1 L. **(a)** Are the two solutions in osmotic equilibrium? **(b)** If not, in which direction will the water flow—from left to right or from right to left? **(c)** What will be the final concentration of Na^+ in each half? **(d)** Is the flow spontaneous or nonspontaneous?

S15-23. Now, what's the difference between a glass vessel and a living cell? Pursue the following analogy: Inside a cell, the concentration of Na^+ has a certain value x. In the water outside the cell, the concentration of Na^+ has the value $x/2$. The intracellular and extracellular regions are separated by a semipermeable membrane through which H_2O molecules (although not sodium ions) can pass. Nevertheless, the concentration of Na^+ inside the cell remains x—unchanged—and the concentration of Na^+ outside the cell remains $x/2$. **(a)** Is the cell in osmotic equilibrium with the external world? If not, explain how the living system can resist nature's powerful tendency to "smooth away the differences." **(b)** Does a living cell violate, in any way, the second law of thermodynamics? If it does, what shall we make of our so-called *law*? If the cell does not violate any law of nature, then how does it manage to conduct its chemical activities legally?

S15-24. What information do you need in order to determine the molar mass of a solute by measurement of osmotic pressure?

S15-25. **(a)** Does the osmotic pressure of a solution increase, decrease, or remain the same as the temperature is increased? **(b)** If solution 1 is isotonic with solution 2 at a certain temperature T, do the two solutions stay isotonic as the temperature is increased? **(c)** What happens if solution 1 (originally isotonic with solution 2) loses half its volume owing to evaporation? Do the two solutions remain isotonic? If not, does solution 1 become hypotonic or hypertonic relative to solution 2?

S15-26. An aqueous solution of sodium chloride (call it solution 1) contains 2.000 g NaCl in a total volume of 100.0 mL. **(a)** How many grams of $Ce_2(SO_4)_3$ must be dissolved in a total volume of 200.0 mL to produce a second solution (call it solution 2) isotonic with solution 1 at 5°C? **(b)** Which piece of information given in part (a) is superfluous to this exercise?

S15-27. A mixture of nitrogen and oxygen gases is in contact with liquid water at 20°C. Henry's law constants are given below:

$$N_2(aq) \rightleftarrows N_2(g) \qquad K(20°C) = 1.4 \times 10^3 \text{ atm L mol}^{-1}$$
$$O_2(aq) \rightleftarrows O_2(g) \qquad K(20°C) = 7.2 \times 10^2 \text{ atm L mol}^{-1}$$

The partial pressure of N_2 is 50 atm, and the partial pressure of O_2 is 10 atm. **(a)** Calculate the solubility of N_2 under the stated conditions. **(b)** Calculate the solubility of N_2 after the partial pressure of O_2 is increased to 100 atm.

S15-28. $AgNO_3$ is strongly soluble in water. AgI is nearly insoluble, with a K_{sp} less than 10^{-16}. At equilibrium, state whether the residual free energy (ΔG) for the dissolution of silver nitrate,

$$AgNO_3(s) \rightleftarrows Ag^+(aq) + NO_3^-(aq)$$

is greater than, less than, or equal to the residual free energy for the dissolution of silver iodide:

$$AgI(s) \rightleftarrows Ag^+(aq) + I^-(aq)$$

Be sure to distinguish ΔG from $\Delta G°$.

SOLUTIONS

S15-1. From the Glossary:

Solubility. The concentration of solute in a saturated solution—thus the maximum amount of solute that can be dissolved in a specified volume of solvent at a specified temperature.

Solubility-product constant. The equilibrium constant for a dissolution–precipitation reaction, usually expressed as a product of ion concentrations.

S15-2. The problem provides us with information only about the kinetics of dissolution for X and Y, not the thermodynamics. It tells us *how fast* the reactions proceed, but it does not tell us ultimately *how much* of each substance can be dissolved.

(a) There is no thermodynamic basis to decide whether the dissolutions are exothermic or endothermic.

(b) Since the solutions are apparently unsaturated (no excess solute is present), we assume that the reactions have not yet come to equilibrium. Presumably, more X and more Y can still be dissolved at each temperature—but how much? We simply cannot tell from the information given. We can say nothing about the relative solubilities of X and Y at the specified temperatures.

(c) Similarly, the information available is insufficient to determine the equilibrium constants.

S15-3. We are given formation data for a hypothetical substance X:

	ΔH_f° (kJ mol^{-1})	ΔG_f° (kJ mol^{-1})
X(s)	120	100
X(aq)	140	80

(a) Since the enthalpy of solution is positive,

$$\text{X(s)} \rightleftarrows \text{X(aq)}$$

$$\Delta H^\circ_{\text{soln}} = \Delta H_f^\circ[\text{X(aq)}] - \Delta H_f^\circ[\text{X(s)}] = (140 - 120)\text{ kJ mol}^{-1} = 20\text{ kJ mol}^{-1}$$

we expect that the equilibrium constant—and hence the solubility—will increase with temperature:

$$K(T) = \exp\left(-\frac{\Delta G^\circ}{RT}\right) = \exp\left(-\frac{\Delta H^\circ}{RT}\right)\exp\left(\frac{\Delta S^\circ}{R}\right)$$

$$\frac{K(T_2)}{K(T_1)} = \exp\left[-\frac{\Delta H^\circ}{R}\left(\frac{1}{T_2} - \frac{1}{T_1}\right)\right] > 1 \quad \text{if} \quad \Delta H^\circ > 0 \text{ and } T_2 > T_1$$

(b) The information given in the problem pertains only to the thermodynamics of the reaction, not the kinetics. We have no basis to predict the speed at which equilibrium will be attained.

S15-4. Our generic ionic compound dissociates into A^{p+} cations and B^{q-} anions:

$$A_jB_k(s) \rightleftarrows jA^{p+}(aq) + kB^{q-}(aq)$$

(a) We require that

$$jp - kq = 0$$

in order to guarantee overall electrical neutrality:

$$p = \frac{k}{j}q$$

(b) The solubility-product constant takes its usual form:

$$K_{sp} = [A^{p+}]^j[B^{q-}]^k$$

The concentration of solid A_jB_k does not appear in the expression.

(c) First, represent the equilibrium concentrations in terms of a single unknown (x):

$$A_jB_k(s) \rightleftarrows jA^{p+}(aq) + kB^{q-}(aq)$$

	A^{p+}	B^{q-}
Initial concentration	0	0
Change	jx	kx
Equilibrium concentration	jx	kx

Second, set up and solve the mass-action equation:

$$K_{sp} = [A^{p+}]^j[B^{q-}]^k = (jx)^j(kx)^k = j^jk^kx^{j+k}$$

$$x^{j+k} = \frac{K_{sp}}{j^jk^k}$$

$$x = \left(\frac{K_{sp}}{j^jk^k}\right)^{\frac{1}{j+k}}$$

Evaluating the expression for x, we have the molar solubility of A_jB_k.

S15-5. Here we apply the results of Exercise S15-4 to the generic salts AB_2 and AB_3. The value of K_{sp} is assumed to be the same for each.

(a) Molar solubilities (x) are calculated below for AB_2 and AB_3:

$$\begin{array}{cc} AB_2 & AB_3 \\ x(2x)^2 = K_{sp} & x(3x)^3 = K_{sp} \\ x = \left(\frac{K_{sp}}{4}\right)^{1/3} & x = \left(\frac{K_{sp}}{27}\right)^{1/4} \end{array}$$

The ratio varies with the magnitude of K_{sp}:

$$\frac{x(AB_2)}{x(AB_3)} = \frac{\left(\frac{K_{sp}}{4}\right)^{1/3}}{\left(\frac{K_{sp}}{27}\right)^{1/4}} = \frac{27^{1/4}}{4^{1/3}}\left(K_{sp}\right)^{1/12} \approx 1.436\left(K_{sp}\right)^{1/12}$$

(b) See above. Cation concentrations, given by x, are determined numerically by K_{sp}. The two values are generally different (except in the special case where $K_{sp} = 4^4/27^3$):

$$\begin{array}{cc} AB_2 & AB_3 \\ x = \left(\frac{K_{sp}}{4}\right)^{1/3} & x = \left(\frac{K_{sp}}{27}\right)^{1/4} \end{array}$$

(c) The concentration of anions is $2x$ in AB_2 and $3x$ in AB_3:

$$\begin{array}{cc} AB_2 & AB_3 \\ 2x = 2\left(\frac{K_{sp}}{4}\right)^{1/3} & 3x = 3\left(\frac{K_{sp}}{27}\right)^{1/4} \end{array}$$

S15-6. $Mg(OH)_2$ is a salt of the type AB_2:

$$\begin{array}{cccc} Mg(OH)_2(s) \rightleftarrows & Mg^{2+}(aq) & + & 2OH^-(aq) \\ & x & & 2x \end{array}$$

$$K_{sp} = [Mg^{2+}][OH^-]^2 = x(2x)^2 = 4x^3$$

Recognize that x denotes the molar solubility, and then determine the equilibrium constant by straightforward conversion of grams into moles:

$$K_{sp} = 4x^3 = 4\left(\frac{0.0065\text{ g Mg(OH)}_2}{\text{L}} \times \frac{1\text{ mol}}{58.3197\text{ g Mg(OH)}_2}\right)^3 = 5.5 \times 10^{-12}$$

In keeping with our usual practice, we treat K_{sp} as a dimensionless number.

S15-7. Another salt of the type AB_2:

$$\begin{array}{cccc} SrF_2(s) \rightleftarrows & Sr^{2+}(aq) & + & 2F^-(aq) \\ & x & & 2x \end{array}$$

$$K_{sp} = [Sr^{2+}][F^-]^2 = x(2x)^2 = 4x^3$$

$$x = \left(\frac{K_{sp}}{4}\right)^{1/3} = \left(\frac{4.3 \times 10^{-9}}{4}\right)^{1/3} = 1.02 \times 10^{-3}\ M$$

Given the molar mass, we express the value in grams per liter:

$$\frac{1.02 \times 10^{-3}\text{ mol SrF}_2}{\text{L}} \times \frac{125.62\text{ g SrF}_2}{\text{mol SrF}_2} = 0.13\text{ g SrF}_2\text{ L}^{-1} \quad \text{(2 sig fig)}$$

S15-8. Use K_{sp} (Table C-19) to calculate the molar solubility of each salt, as noted below:

Hg_2I_2	<	$ZnCO_3$	<	MgF_2	<	$Ca(NO_3)_2$
$2.4 \times 10^{-10}\ M$		$1.1 \times 10^{-5}\ M$		$1.2 \times 10^{-3}\ M$		soluble

S15-9. Silver chloride serves as our prototype AB salt:

$$\begin{array}{cccc} AgCl(s) \rightleftarrows & Ag^+(aq) & + & Cl^-(aq) \\ & x & & x \end{array}$$

$$K_{sp} = [Ag^+][Cl^-] = x^2$$

$$x = \sqrt{K_{sp}} = \sqrt{1.8 \times 10^{-10}} = 1.34 \times 10^{-5}\ M$$

We retain, for the moment, one extra digit in the molar solubility x.

(a) $$\frac{1\text{ L}}{1.34 \times 10^{-5}\text{ mol AgCl}} \times \frac{1\text{ mol AgCl}}{143.321\text{ g AgCl}} \times 1.0\text{ g AgCl} = 5.2 \times 10^2\text{ L}$$

(b) $$\frac{1\text{ L}}{1.34 \times 10^{-5}\text{ mol AgCl}} \times \frac{1\text{ mol AgCl}}{143.321\text{ g AgCl}} \times \left(1.0 \times 10^2\right)\text{g AgCl} = 5.2 \times 10^4\text{ L}$$

S15-10. Take the same approach as in Exercise S15-9, this time with $PbCl_2$ (an AB_2 salt):

$$PbCl_2(s) \rightleftarrows Pb^{2+}(aq) + 2Cl^-(aq)$$
$$\qquad\qquad x \qquad\qquad 2x$$

$$K_{sp} = [Pb^{2+}][Cl^-]^2 = x(2x)^2 = 4x^3$$

$$x = \left(\frac{K_{sp}}{4}\right)^{1/3} = \left(\frac{1.6 \times 10^{-5}}{4}\right)^{1/3} = 0.0159\ M$$

(a) $$\frac{1\ \text{L}}{0.0159\ \text{mol PbCl}_2} \times \frac{1\ \text{mol PbCl}_2}{278.1\ \text{g PbCl}_2} \times 1.0\ \text{g PbCl}_2 = 0.23\ \text{L}$$

(b) $$\frac{1\ \text{L}}{0.0159\ \text{mol PbCl}_2} \times \frac{1\ \text{mol PbCl}_2}{278.1\ \text{g PbCl}_2} \times \left(1.0 \times 10^2\right)\text{g PbCl}_2 = 23\ \text{L}$$

S15-11. The solubility of $MgCO_3$ is the same both in pure water and in a solution containing Na_2SO_4. Sodium and sulfate ions do not participate in the heterogeneous equilibrium between $MgCO_3(s)$ and $MgCO_3(aq)$. There are no common-ion effects, and no new compounds are precipitated. Na_2CO_3 and $MgSO_4$ are soluble salts (Table C-17).

S15-12. The solubility of $MgCO_3$ in a solution containing 1.00 M $MgSO_4$ will be less than x, the solubility in pure water. The high concentration of Mg^{2+} already present in $MgSO_4(aq)$ will suppress any subsequent dissolution of $MgCO_3$.

S15-13. A reminder about the nature of heterogeneous equilibrium in an ionic solution.

(a) A small amount of additional AgCl(s) will dissolve in the slightly larger volume created by each drop of water.

(b) The concentration of AgCl(aq), fixed by K_{sp}, remains the same. More moles of solute dissolve in a larger volume of solvent.

S15-14. Heterogeneous equilibrium, continued.

(a) The equilibrium constant K_{sp}, a function of temperature alone, has the same value whether the amount of undissolved solute is large or small. As long as there is a single grain of solid material, the solution is saturated. The equilibrium concentration remains equal to its previous value.

(b) The reaction quotient falls below K_{sp} as additional water enters the solution. The grain of solid AgCl dissolves, and the concentration of solute decreases to less than its equilibrium value. Dilution leaves the mixture unsaturated.

(c) The concentration of AgCl(aq) continues to decrease as more water is added.

S15-15. In Exercise S15-14 we considered a nonequilibrium system with too small a concentration of solute. In the present exercise we consider one with too *large* a concentration.

(a) No, the solution is supersaturated—temporarily out of equilibrium.

(b) The reaction quotient is greater than K_{sp} in the supersaturated solution: $Q > K_{sp}$.

(c) The numerical value of K_{sp} depends only on the temperature. It remains the same regardless of concentration and volume.

(d) The equilibrium concentration of solute is the same in both systems. A certain amount of material must precipitate from solution in order to establish equilibrium in the smaller volume.

S15-16. The absolute amount (in moles or grams) is different, but the equilibrium *concentration* (amount per unit volume) is the same before and after evaporation:

$$\text{AgOH(s)} \rightleftarrows \underset{x}{\text{Ag}^+\text{(aq)}} + \underset{x}{\text{OH}^-\text{(aq)}}$$

$$K_{sp} = [\text{Ag}^+][\text{OH}^-] = 1.8 \times 10^{-8}$$

$$x^2 = 1.8 \times 10^{-8}$$

$$x = 1.34 \times 10^{-4}$$

Rounded to two significant figures, we have $[\text{Ag}^+] = [\text{OH}^-] = 1.3 \times 10^{-4}$ M regardless of the total volume of solution.

S15-17. We are given the following hypothetical data:

REACTION	$\Delta H°$ (kJ mol^{-1})	$\Delta S°$ (J mol^{-1} K^{-1})
$\text{AB(s)} \rightleftarrows \text{A}^+\text{(aq)} + \text{B}^-\text{(aq)}$	50	100
$\text{CD(s)} \rightleftarrows \text{C}^+\text{(aq)} + \text{D}^-\text{(aq)}$	–50	100

Note that the dissolution of AB is endothermic, whereas the dissolution of CD is exothermic, The enthalpies of solution are equal in magnitude but opposite in sign. Both compounds, however, share the same entropy of solution.

(a) Calculate ΔG° for each dissolution, and note the relationship between free energy and the equilibrium constant:

$$K = \exp\left(-\frac{\Delta G^\circ}{RT}\right)$$

The more negative the value of ΔG° at a given temperature, the greater will be the value of K (and hence the solubility).

For AB, with a positive free energy of solution, the equilibrium constant is less than 1:

$$\text{AB(s)} \rightleftarrows \text{A}^+\text{(aq)} + \text{B}^-\text{(aq)}$$

$$\Delta G^\circ(\text{AB}) = \Delta H^\circ(\text{AB}) - T\Delta S^\circ(\text{AB})$$

$$= 50 \text{ kJ mol}^{-1} - (298 \text{ K})(0.100 \text{ kJ mol}^{-1} \text{ K}^{-1}) = 20.2 \text{ kJ mol}^{-1}$$

By contrast, the free energy of solution is strongly negative for CD. The equilibrium constant is substantially greater than 1:

$$\text{CD(s)} \rightleftarrows \text{C}^+\text{(aq)} + \text{D}^-\text{(aq)}$$

$$\Delta G^\circ(\text{CD}) = \Delta H^\circ(\text{CD}) - T\Delta S^\circ(\text{CD})$$

$$= -50 \text{ kJ mol}^{-1} - (298 \text{ K})(0.100 \text{ kJ mol}^{-1} \text{ K}^{-1}) = -79.8 \text{ kJ mol}^{-1}$$

Solute CD is more soluble at 25°C.

(b) The enthalpy of solution is positive for AB. The reaction is endothermic, and the equilibrium constant increases with temperature:

$$K(T) = \exp\left(-\frac{\Delta G^\circ}{RT}\right) = \exp\left(-\frac{\Delta H^\circ}{RT}\right)\exp\left(\frac{\Delta S^\circ}{R}\right)$$

$$\frac{K(T_2)}{K(T_1)} = \exp\left[-\frac{\Delta H^\circ}{R}\left(\frac{1}{T_2} - \frac{1}{T_1}\right)\right] > 1 \quad \text{if} \quad \Delta H^\circ > 0 \text{ and } T_2 > T_1$$

Solubility increases accordingly.

(c) The opposite effect: For CD, the enthalpy of solution is negative. The reaction is exothermic, and the equilibrium constant decreases with temperature:

$$\frac{K(T_2)}{K(T_1)} = \exp\left[-\frac{\Delta H^\circ}{R}\left(\frac{1}{T_2} - \frac{1}{T_1}\right)\right] < 1 \quad \text{if} \quad \Delta H^\circ < 0 \quad \text{and} \quad T_2 > T_1$$

Solubility decreases accordingly.

S15-18. Similar to the preceding exercise, but now the entropy of solution is negative for solute CD:

REACTION	ΔH° (kJ mol^{-1})	ΔS° (J mol^{-1} K^{-1})
$AB(s) \rightleftarrows A^+(aq) + B^-(aq)$	50	100
$CD(s) \rightleftarrows C^+(aq) + D^-(aq)$	−50	−300

(a) Compound AB, with the *less positive* value of ΔG°, is more soluble at 298 K. The unfavorable change in entropy impairs the solubility of CD:

$$\Delta G^\circ(\text{AB}) = \Delta H^\circ(\text{AB}) - T\,\Delta S^\circ(\text{AB})$$

$$= 50 \text{ kJ mol}^{-1} - (298\text{ K})(0.100 \text{ kJ mol}^{-1}\text{ K}^{-1}) = 20.2 \text{ kJ mol}^{-1}$$

$$\Delta G^\circ(\text{CD}) = \Delta H^\circ(\text{CD}) - T\,\Delta S^\circ(\text{CD})$$

$$= -50 \text{ kJ mol}^{-1} - (298\text{ K})(-0.300 \text{ kJ mol}^{-1}\text{ K}^{-1}) = 39.4 \text{ kJ mol}^{-1}$$

(b) The solubility of AB increases with temperature because the process is endothermic. See Exercise S15-17.

(c) The opposite effect: The solubility of CD decreases with temperature because the process is exothermic.

S15-19. One more pair of dissolution–precipitation equilibria:

REACTION	ΔH° (kJ mol^{-1})	ΔS° (J mol^{-1} K^{-1})
$AB(s) \rightleftarrows A^+(aq) + B^-(aq)$	50	100
$CD(s) \rightleftarrows C^+(aq) + D^-(aq)$	5	−100

Free energies of solution at 298 K are calculated below:

$$\Delta G^\circ(\text{AB}) = \Delta H^\circ(\text{AB}) - T\Delta S^\circ(\text{AB})$$

$$= 50 \text{ kJ mol}^{-1} - (298 \text{ K})(0.100 \text{ kJ mol}^{-1} \text{ K}^{-1}) = 20.2 \text{ kJ mol}^{-1}$$

$$\Delta G^\circ(\text{CD}) = \Delta H^\circ(\text{CD}) - T\Delta S^\circ(\text{CD})$$

$$= 5 \text{ kJ mol}^{-1} - (298 \text{ K})(-0.100 \text{ kJ mol}^{-1} \text{ K}^{-1}) = 34.8 \text{ kJ mol}^{-1}$$

(a) At 298 K, the equilibrium constant for the dissolution of AB is larger by more than two orders of magnitude:

$$\frac{K_{\text{sp}}(\text{AB})}{K_{\text{sp}}(\text{CD})} = \frac{\exp\left[-\dfrac{\Delta G^\circ(\text{AB})}{RT}\right]}{\exp\left[-\dfrac{\Delta G^\circ(\text{CD})}{RT}\right]} = \exp\left[-\frac{(20.2 - 34.8) \text{ kJ mol}^{-1}}{(8.3145 \times 10^{-3} \text{ kJ mol}^{-1} \text{ K}^{-1})(298 \text{ K})}\right]$$

$$= 3.6 \times 10^2$$

(b) Reevaluate ΔG° at 277 K,

$$\Delta G^\circ(\text{AB}) = \Delta H^\circ(\text{AB}) - T\Delta S^\circ(\text{AB})$$

$$= 50 \text{ kJ mol}^{-1} - (277 \text{ K})(0.100 \text{ kJ mol}^{-1} \text{ K}^{-1}) = 22.3 \text{ kJ mol}^{-1}$$

$$\Delta G^\circ(\text{CD}) = \Delta H^\circ(\text{CD}) - T\Delta S^\circ(\text{CD})$$

$$= 5 \text{ kJ mol}^{-1} - (277 \text{ K})(-0.100 \text{ kJ mol}^{-1} \text{ K}^{-1}) = 32.7 \text{ kJ mol}^{-1}$$

and recompute the ratio of equilibrium constants:

$$\frac{K_{\text{sp}}(\text{AB})}{K_{\text{sp}}(\text{CD})} = \frac{\exp\left[-\dfrac{\Delta G^\circ(\text{AB})}{RT}\right]}{\exp\left[-\dfrac{\Delta G^\circ(\text{CD})}{RT}\right]} = \exp\left[-\frac{(22.3 - 32.7) \text{ kJ mol}^{-1}}{(8.3145 \times 10^{-3} \text{ kJ mol}^{-1} \text{ K}^{-1})(277 \text{ K})}\right] = 91$$

(c) The ratio improves in favor of CD as the temperature decreases. The unfavorable change in entropy (-100 J mol^{-1} K^{-1}) is outweighed by the smaller endothermicity (5 kJ mol^{-1}) compared with AB.

S15-20. First, calculate the standard change in free energy:

$$\text{Li}_2\text{CO}_3(\text{s}) \rightleftarrows 2\text{Li}^+(\text{aq}) + \text{CO}_3^{2-}(\text{aq})$$

$$\begin{aligned}\Delta G^\circ &= 2\,\Delta G_\text{f}^\circ\left[\text{Li}^+(\text{aq})\right] + \Delta G_\text{f}^\circ\left[\text{CO}_3^{2-}(\text{aq})\right] - \Delta G_\text{f}^\circ\left[\text{Li}_2\text{CO}_3(\text{s})\right] \\ &= \left(-\frac{293.3\text{ kJ}}{\text{mol}} \times 2\text{ mol}\right) + \left(-\frac{527.8\text{ kJ}}{\text{mol}} \times 1\text{ mol}\right) - \left(-\frac{1132.1\text{ kJ}}{\text{mol}} \times 1\text{ mol}\right) \\ &= 17.7\text{ kJ} \quad (\text{per mole of Li}_2\text{CO}_3)\end{aligned}$$

Then insert ΔG° into the exponential equation for the equilibrium constant:

$$K_\text{sp} = \exp\left(-\frac{\Delta G^\circ}{RT}\right) = \exp\left[-\frac{17.7\text{ kJ mol}^{-1}}{\left(8.3145 \times 10^{-3}\text{ kJ mol}^{-1}\text{ K}^{-1}\right)(298\text{ K})}\right] = 7.9 \times 10^{-4}$$

S15-21. Tables C-17 and C-18 in the text show that $Ca(NO_3)_2$ is strongly soluble in water. Its solubility-product constant, K_{sp}, should be greater than 1.

S15-22. We begin a series of five exercises dealing with osmosis and osmotic pressure.

(a) Initially the two halves of the container are not in osmotic equilibrium. The concentration of Na^+ is different on each side: x on the left, $x/2$ on the right.

(b) Water will flow from right to left—from the less concentrated solution to the more concentrated solution. The concentration will decrease on the left and increase on the right.

(c) The nonequilibrium system is initially in the following condition:

LEFT	RIGHT
x mol	x/2 mol
1 L	1 L

A flow of water (volume = ΔV) subsequently equalizes the concentrations,

$$\frac{x\text{ mol}}{(1+\Delta V)\text{ L}} = \frac{x/2\text{ mol}}{(1-\Delta V)\text{ L}}$$

and we solve the resulting linear equation for ΔV:

$$\frac{1+\Delta V}{2} = 1 - \Delta V$$

$$\Delta V = \frac{1}{3}$$

The final concentrations ($\frac{3}{4}x$ mol L^{-1}) are equal on both sides:

LEFT	RIGHT
x mol 4/3 L	x/2 mol 2/3 L

(d) The flow is spontaneous. It erases an imbalance in free energy.

S15-23. Osmosis, continued.

(a) No, the cell is not in osmotic equilibrium with the world outside. It must expend energy to prevent an equalization of concentrations.

(b) The cell does not violate the second law of thermodynamics. It derives its energy from spontaneous reactions and uses part of the energetic proceeds to fund its nonspontaneous activities. The net result is an increase in global entropy.

S15-24. The equation for osmotic pressure is

$$\Pi = \frac{n}{V}RT = \frac{mRT}{\mathcal{m}V}$$

where $\mathcal{m}$ is the molar mass of the solute. To determine $\mathcal{m}$ we need values for the other four quantities: Π, the osmotic pressure; m, the macroscopic mass in grams; T, the absolute temperature; V, the volume.

S15-25. Osmotic pressure depends on the concentration ($c = n/V$) and absolute temperature (T):

$$\Pi = \frac{n}{V}RT = cRT$$

(a) Osmotic pressure increases with temperature, as evident in the equation above.

(b) As long as the concentrations stay the same, the two solutions will remain isotonic at the higher temperature.

(c) If solution 1 loses half its volume, then both its concentration and osmotic pressure are doubled relative to solution 2. Solution 1 becomes hypertonic with respect to solution 2.

S15-26. Osmotic pressure, a colligative property, depends on the *total* concentration of dissolved particles.

(a) Calculate, first, the concentration of dissolved ions in the solution of NaCl:

$$\frac{2.000 \text{ g NaCl}}{100.0 \text{ mL}} \times \frac{1 \text{ mol NaCl}}{58.443 \text{ g NaCl}} \times \frac{2 \text{ mol ions}}{\text{mol NaCl}} = \frac{6.844 \times 10^{-4} \text{ mol ions}}{\text{mL}}$$

Then calculate the equivalent mass of cerium(III) sulfate:

$$\frac{6.844 \times 10^{-4} \text{ mol ions}}{\text{mL}} \times \frac{1 \text{ mol Ce}_2(\text{SO}_4)_3}{5 \text{ mol ions}} \times \frac{568.423 \text{ g Ce}_2(\text{SO}_4)_3}{\text{mol Ce}_2(\text{SO}_4)_3} \times 200.0 \text{ mL}$$

$$= 15.56 \text{ g Ce}_2(\text{SO}_4)_3$$

(b) The temperature is irrelevant in this particular example. We only need to compute the concentration of dissolved ions.

S15-27. We are given Henry's law constants for molecular nitrogen and oxygen:

$$N_2(aq) \rightleftarrows N_2(g) \qquad K(20°C) = 1.4 \times 10^3 \text{ atm L mol}^{-1}$$

$$O_2(aq) \rightleftarrows O_2(g) \qquad K(20°C) = 7.2 \times 10^2 \text{ atm L mol}^{-1}$$

(a) Substitute the partial pressure of $N_2(g)$ into the equilibrium equation, and solve for the dissolved concentration $[N_2(aq)]$:

$$K = \frac{P_{N_2}}{[N_2(aq)]}$$

$$[N_2(aq)] = \frac{P_{N_2}}{K} = \frac{50 \text{ atm}}{1.4 \times 10^3 \text{ atm L mol}^{-1}} = 0.036 \text{ mol L}^{-1}$$

(b) The solubility of N_2 remains the same—0.036 *M*—because the equilibrium between $N_2(g)$ and $N_2(aq)$ is unaffected by the partial pressure of O_2.

S15-28. The residual free energy, ΔG, is *zero* for all reactions at equilibrium, no matter what the value of the equilibrium constant may be. The free energy of the products is equal to the free energy of the reactants. There is no drive to produce one species at the expense of another.

Chapter 16

Acids and Bases

S16-1. Listed below are four statements concerning an aqueous solution of HCl at 25°C. Which one is true?

(a) The pH is always less than 7.
(b) The pH is always greater than 7.
(c) The pH is always greater than 0.
(d) The pH is usually 1.

S16-2. An aqueous solution contains 0.000001 mol HCl. Which of the following statements is true?

(a) The pH of the solution is 6.
(b) The pH of the solution is –6.
(c) The pH of the solution lies between 6 and 7.
(d) There is insufficient information to determine the pH.

S16-3. Beaker 1 contains a solution of hydrochloric acid in water. Beaker 2 contains a solution of acetic acid in water, the same volume as in beaker 1. Which of the following statements is true?

(a) The pH of solution 1 is always higher than the pH of solution 2.
(b) The pH of solution 1 is always lower than the pH of solution 2.
(c) The pH of solution 1 is always equal to the pH of solution 2.
(d) The pH of solution 1 may be higher than, lower than, or equal to the pH of solution 2.

S16-4. The initial concentration of HCl in a certain solution is 10^{-8} *M*. Which of the following statements is true?

(a) The pH at 25°C is exactly 6.
(b) The pH at 25°C is slightly less than 7.
(c) The pH at 25°C is exactly 7.
(d) The pH at 25°C is slightly greater than 7.
(e) The pH at 25°C is exactly 8.

S16-5. The equilibrium constant for the autoionization of water is 1.0×10^{-14} at 25°C:

$$H_2O(\ell) + H_2O(\ell) \rightleftarrows H_3O^+(aq) + OH^-(aq) \qquad K_w(25°C) = 1.0 \times 10^{-14}$$

Which of the following statements is true for a strongly *acidic* solution at 25°C?

(a) The numerical value of K_w is greater than 10^{-14}.
(b) The numerical value of K_w is less than 10^{-14}.
(c) The numerical value of K_w is equal to 10^{-14}.
(d) There is insufficient information to determine the value of K_w in the presence of acid.

S16-6. Consider again the autoionization of water:

$$H_2O(\ell) + H_2O(\ell) \rightleftarrows H_3O^+(aq) + OH^-(aq)$$

Which of the following statements is true?

(a) The equilibrium shifts to the left in the presence of strong acid.
(b) The equilibrium shifts to the right in the presence of strong acid.
(c) The equilibrium is unaffected by the presence of strong acid.
(d) There is insufficient information to determine the effect of strong acid on the equilibrium.

S16-7. The pH of pure water at 10°C is 7.26. **(a)** Determine the value of K_w at this temperature. **(b)** Is water acidic, basic, or neutral at 10°C?

S16-8. **(a)** Use the appropriate data in Appendix C to compute the pH of pure water at 50°C. **(b)** Compute the corresponding pOH. **(c)** Is water acidic, basic, or neutral at 50°C?

S16-9. Is it possible for an aqueous solution of NaOH at 25°C to have a pH less than 7? If yes, give an example of such a solution. If no, explain why not.

S16-10. An aqueous solution contains 45.0 g NaCl dissolved in a total volume of 450.0 mL. **(a)** Calculate the pH at 25°C. **(b)** What happens to the pH when the solution is diluted by a factor of 10?

S16-11. For each of the following salts, write a neutralization reaction able to produce the given compound:

(a) KIO_4 **(b)** $Ca(HCOO)_2$ **(c)** NH_4CH_3COO **(d)** BaI_2

S16-12. Rank the following acids in order of increasing strength:

$$H_3AsO_4, \ H_2CO_3, \ HClO_3, \ CF_3COOH$$

Consult Appendix C as needed.

S16-13. Rank the following conjugate bases in order of increasing strength:

$$HCOO^-, \ F^-, \ BrO^-, \ SCN^-$$

Consult Appendix C as needed.

S16-14. For each of the following salts, state whether the pH of an aqueous solution at 25°C is greater than, less than, or equal to 7—or, perhaps, impossible to establish without further information:

(a) $NaCH_3CH_2COO$ **(b)** KBr **(c)** NH_4Cl **(d)** $NaNO_3$

Consult Appendix C as needed.

S16-15. Bromoacetic acid ($CH_2BrCOOH$) has a pK_a value of 2.70. Given an initial concentration $[CH_2BrCOOH]_0$, calculate the pH and the ratio $[CH_2BrCOO^-]/[CH_2BrCOOH]$ at equilibrium:

(a) $[CH_2BrCOOH]_0 = 1.000\ M$
(b) $[CH_2BrCOOH]_0 = 0.500\ M$
(c) $[CH_2BrCOOH]_0 = 0.100\ M$
(d) $[CH_2BrCOOH]_0 = 0.010\ M$

S16-16. A buffer solution is prepared as a mixture of bromoacetic acid ($pK_a = 2.70$) and its conjugate base. The pH is 3.00. **(a)** In what proportions must the two components be mixed in order to realize the stated pH? **(b)** Suppose that a certain amount of NaOH is added to the buffer solution. Do you have sufficient information to determine the change in pH? If yes, do so. If no, what further information do you need?

S16-17. A mixture contains 0.00100 M $CH_2BrCOOH$ and 0.00100 M CH_2BrCOO^-. The ionization constant of bromoacetic acid is $K_a(25°C) = 0.0020$. **(a)** Do you expect this system to act as an effective buffer? **(b)** Calculate the pH. Is it appropriate to apply the Henderson-Hasselbalch equation here?

S16-18. Given an initial concentration of formic acid, $[HCOOH]_0$, calculate the pH and the ratio $[HCOO^-]/[HCOOH]$ at equilibrium:

(a) $[HCOOH]_0 = 1.000\ M$

(b) $[HCOOH]_0 = 0.500\ M$

(c) $[HCOOH]_0 = 0.100\ M$

(d) $[HCOOH]_0 = 0.010\ M$

The ionization constant of HCOOH is $K_a(25°C) = 1.8 \times 10^{-4}$.

S16-19. A buffer solution is prepared from formic acid (HCOOH) and sodium formate (NaHCOO). The concentration of acid is 1.000 *M* in a volume of 1.000 L, and the pH of the solution is 3.80. **(a)** Calculate the formate concentration. **(b)** Calculate the pH of the buffer solution after addition of 0.100 mol NaOH. **(c)** Recalculate the pH of the original buffer solution after addition of 0.100 mol HCl.

S16-20. A buffer solution contains 5.75 g HCOOH and 6.80 g NaHCOO dissolved in a total volume of 250. mL. **(a)** Calculate the pH. **(b)** What concentration of HCOOH will adjust the pH to 3.50? **(c)** What concentration of HCl, if added to the original buffer, will likewise adjust the pH to 3.50? Assume that the volume does not change.

S16-21. A buffer solution contains 1.000 mol HA and 1.000 mol A^- in a total volume of 1.000 L. Assume that the ionization constant for the reaction

$$HA(aq) + H_2O(\ell) \rightleftarrows A^-(aq) + H_3O^+(aq)$$

has the value $K_a = 1.00 \times 10^{-4}$. **(a)** Calculate the pH. **(b)** Recalculate the pH after addition of 0.001 mol HCl. **(c)** Increase the volume of the original buffer solution by a factor of 100, from 1.000 L to 100.000 L. Recalculate the pH. **(d)** Add 0.1 mol HCl to the diluted buffer solution described in part (c). What is the pH now?

S16-22. Suppose that the ionization reaction of HA is exothermic. Will the pH of an aqueous solution of HA increase, decrease, or remain the same as the temperature is increased?

S16-23. Will the pH of an aqueous solution of HA increase, decrease, or remain the same as the pressure is increased?

S16-24. Will the pH of an aqueous solution of HA increase, decrease, or remain the same when a given amount of HA is dissolved in a larger volume of solvent?

S16-25. HNO_3 is a strong acid. HCN is a weak acid. At equilibrium, state whether the residual free energy (ΔG) for the ionization of nitric acid,

$$HNO_3(aq) + H_2O(\ell) \rightleftarrows NO_3^-(aq) + H_3O^+(aq)$$

is greater than, less than, or equal to the residual free energy for the ionization of hydrocyanic acid:

$$HCN(aq) + H_2O(\ell) \rightleftarrows CN^-(aq) + H_3O^+(aq)$$

Be sure to distinguish ΔG from $\Delta G°$.

SOLUTIONS

S16-1. The correct choice is (a). The pH is always less than 7, since even the slightest excess of acid will increase the concentration of H_3O^+ to greater than 10^{-7} *M*.

S16-2. The correct choice is (d). Not knowing the total volume of solution, we have no way to determine the concentration of H_3O^+.

S16-3. The correct choice is (d). The pH of each solution is determined by the initial concentration of acid, which is unknown. Hence we have no way to decide whether the pH of solution 1 is higher or lower than the pH of solution 2.

For example, a highly dilute solution of HCl may have a higher pH than a concentrated solution of acetic acid. Any relationship is possible.

S16-4. The correct choice is (b). The pH of a 10^{-8} *M* solution of HCl is slightly less than 7—and is *not* equal to 8, despite what the number 10^{-8} seems to suggest. Water itself is the dominant source of H_3O^+ here, producing a 10^{-7} *M* concentration as a result of autoionization. A small number of H_3O^+ ions supplied by hydrochloric acid causes $[H_3O^+]$ to increase slightly above 10^{-7} *M*. In response, the pH falls to just below 7.

S16-5. The correct choice is (c): $K_w(25°C) = 1.0 \times 10^{-14}$. The value of the equilibrium constant is determined by the temperature alone, not by the concentration of a particular species involved in the equilibrium.

S16-6. The correct choice is (a). The equilibrium shifts to the left in the presence of strong acid, a consequence of Le Châtelier's principle. H_3O^+ supplied by the acid, indistinguishable from the autoionization product H_3O^+, acts as a common ion.

S16-7. The autoionization constant for water, like other equilibrium constants, is a function of temperature.

(a) The stoichiometry of the autoionization reaction ensures that H_3O^+ and OH^- are produced in equal concentrations:

$$2H_2O(\ell) \rightleftarrows \underset{x}{H_3O^+(aq)} + \underset{x}{OH^-(aq)}$$

Knowing the pH, we simply calculate

$$x = [H_3O^+] = 10^{-pH} = 10^{-7.26}$$

and insert the value into K_w:

$$K_w(10°C) = [H_3O^+][OH^-] = x^2 = 10^{-14.52} = 3.0 \times 10^{-15}$$

(b) The system is neutral. There are equal concentrations of H_3O^+ and OH^-.

S16-8. We continue to treat the autoionization of water:

$$2H_2O(\ell) \rightleftarrows \underset{x}{H_3O^+(aq)} + \underset{x}{OH^-(aq)}$$

(a) Table C-15 in *Principles of Chemistry* gives us the value $K_w(50°C) = 5.3 \times 10^{-14}$:

$$[H_3O^+][OH^-] = K_w$$

$$x^2 = 5.3 \times 10^{-14}$$

$$x = 2.3 \times 10^{-7}$$

The corresponding pH is 6.64:

$$pH = -\log[H_3O^+] = -\log\left(2.3 \times 10^{-7}\right) = 6.64$$

(b) Since the autoionization of water produces equal concentrations of H_3O^+ and OH^-, the pOH is necessarily equal to the pH:

$$[OH^-] = [H_3O^+] = x = 2.3 \times 10^{-7}$$

$$pOH = -\log[OH^-] = 6.64$$

(c) The system is neutral. Concentrations of H_3O^+ and OH^- are still equal, even though each of the values is higher than it is at 25°C.

S16-9. No, a solution of NaOH at 25°C will always have a pH greater than 7. The slightest excess of hydroxide ion will increase $[OH^-]$ to a level greater than 10^{-7} *M*.

S16-10. No calculation is necessary.

(a) The pH is 7, because NaCl is the salt of a strong acid (HCl) and a strong base (NaOH). All the hydronium and hydroxide ions present in solution come from the autoionization of water.

(b) Again, a solution of NaCl at 25°C—in whatever amount and in whatever volume—has a pH equal to 7.

S16-11. The anion is contributed by the acid. The cation is contributed by the base.

(a) $HIO_4 + KOH \rightarrow KIO_4 + H_2O$

(b) $2HCOOH + Ca(OH)_2 \rightarrow Ca(HCOO)_2 + 2H_2O$

(c) $CH_3COOH + NH_4OH \rightarrow NH_4CH_3COO + H_2O$

(d) $2HI + Ba(OH)_2 \rightarrow BaI_2 + 2H_2O$

S16-12. See Table C-20 in the text. The lower the value of pK_a, the stronger the acid:

	H_2CO_3 <	H_3AsO_4 <	CF_3COOH <	$HClO_3$
pK_a	6.37	2.26	0.23	<0

S16-13. First, rank the corresponding *acids* in order of increasing strength:

	HBrO <	HCOOH <	HF <	HSCN
pK_a	8.55	3.75	3.17	<0

The strengths of the conjugate bases increase in the reverse order:

$$SCN^- < F^- < HCOO^- < BrO^-$$

S16-14. We need to identify the parentage of each salt: the base responsible for its cation, and the acid responsible for its anion.

(a) $NaCH_3CH_2COO$ is the salt of a weak acid (CH_3CH_2COOH, $pK_a = 4.86$) and a strong base (NaOH). Hydrolysis of the conjugate base produces excess hydroxide ion:

$$CH_3CH_2COO^-(aq) + H_2O(\ell) \rightleftarrows CH_3CH_2COOH(aq) + OH^-(aq)$$

The resulting solution is basic, with pH greater than 7 at 25°C.

(b) KBr is the salt of a strong acid (HBr) and a strong base (KOH). A solution of potassium bromide is neutral: pH = 7.

(c) NH_4Cl is the salt of a strong acid (HCl) and a weak base (NH_3). Hydrolysis of the conjugate acid yields an acidic solution:

$$NH_4^+(aq) + H_2O(\ell) \rightleftarrows NH_3(aq) + H_3O^+(aq)$$

The pH at equilibrium is less than 7.

(d) $NaNO_3$ is the salt of a strong acid (HNO_3) and a strong base (NaOH). A solution of sodium nitrate is neutral: pH = 7.

S16-15. One sample calculation, for $[CH_2BrCOOH]_0 = 1.000\ M$, will be enough to demonstrate the general technique:

$$CH_2BrCOOH(aq) + H_2O(\ell) \rightleftarrows CH_2BrCOO^-(aq) + H_3O^+(aq)$$

	$CH_2BrCOOH$	CH_2BrCOO^-	H_3O^+
Initial conc.	1.000	0	0
Change	$-x$	x	x
Equilibrium conc.	$1.000 - x$	x	x

The mass-action expression leads to an equation quadratic in x:

$$\frac{[H_3O^+][CH_2BrCOO^-]}{[CH_2BrCOOH]} = K_a = 10^{-pK_a} = 10^{-2.70} = 0.0020$$

$$\frac{x^2}{1.000 - x} = 0.0020$$

Solution of the full quadratic equation yields $x = 0.044$ as the physically valid root:

$$0 = x^2 + 0.0020x - 0.0020$$

$$x = \frac{-0.0020 \pm \sqrt{(0.0020)^2 - 4(1)(-0.0020)}}{2(1)} = 0.0437,\ -0.0457$$

With x in hand, we then calculate the pH and the anion-to-acid ratio at equilibrium:

$$\text{pH} = -\log[\text{H}_3\text{O}^+] = -\log x = 1.36$$

$$\frac{[\text{CH}_2\text{BrCOO}^-]}{[\text{CH}_2\text{BrCOOH}]} = \frac{0.0437}{1.000 - 0.0437} = 0.046$$

The remaining initial concentrations are handled in the same way.

	$[\text{CH}_2\text{BrCOOH}]_0$ (M)	pH	$\frac{[\text{CH}_2\text{BrCOO}^-]}{[\text{CH}_2\text{BrCOOH}]}$
(a)	1.000	1.36	0.046
(b)	0.500	1.51	0.065
(c)	0.100	1.88	0.15
(d)	0.010	2.45	0.56

S16-16. We apply the Henderson-Hasselbalch equation to a buffer solution composed of $\text{CH}_2\text{BrCOO}^-$ and CH_2BrCOOH:

$$\text{pH} = \text{p}K_\text{a} + \log\frac{[\text{CH}_2\text{BrCOO}^-]_0}{[\text{CH}_2\text{BrCOOH}]_0}$$

(a) Insert the values given for pH and pK_a, and solve for the ratio of anion to acid:

$$\log\frac{[\text{CH}_2\text{BrCOO}^-]_0}{[\text{CH}_2\text{BrCOOH}]_0} = \text{pH} - \text{p}K_\text{a} = 3.00 - 2.70 = 0.30$$

$$\frac{[\text{CH}_2\text{BrCOO}^-]_0}{[\text{CH}_2\text{BrCOOH}]_0} = 10^{0.30} = 2.0$$

(b) The information provided is insufficient to compute the change in pH. We need to know the absolute concentrations $[\text{CH}_2\text{BrCOO}^-]_0$ and $[\text{CH}_2\text{BrCOOH}]_0$, not just their ratio:

$$\text{pH} = \text{p}K_\text{a} + \log\frac{[\text{CH}_2\text{BrCOO}^-]_0 + \delta}{[\text{CH}_2\text{BrCOOH}]_0 - \delta}$$

S16-17. To be effective, a buffer must maintain ample reserves of both the weak acid and its conjugate base.

(a) $\text{CH}_2\text{BrCOO}^-$ and CH_2BrCOOH are not present in quantities large enough to act as a good buffer.

(b) First, represent the equilibrium concentrations of all three species in terms of a single unknown x:

	$CH_2BrCOOH(aq)$ + $H_2O(\ell)$ $\rightleftarrows$	$CH_2BrCOO^-(aq)$ +	$H_3O^+(aq)$
Initial conc.	0.00100	0.00100	0
Change	$-x$	x	x
Equilibrium conc.	$0.00100 - x$	$0.00100 + x$	x

Next, set up the mass-action expression:

$$K_a = \frac{[H_3O^+][CH_2BrCOO^-]}{[CH_2BrCOOH]} = \frac{x(0.00100 + x)}{(0.00100 - x)} = 0.0020$$

Using the Henderson-Hasselbalch equation here is tantamount to assuming that

$$0.00100 \pm x \approx 0.00100$$

and solving the simplified equation that results:

$$\frac{x(0.00100)}{0.00100} = 0.0020$$

$$x = 0.0020$$

The assumption is clearly bad, however, and the number obtained (pH = pK_a = 2.70) is highly inaccurate. Solving instead the full quadratic equation,

$$x^2 + 0.0030x - (2.0 \times 10^{-6}) = 0$$

we determine that $x = [H_3O^+] = 0.00056$:

$$x = \frac{-0.00300 \pm \sqrt{(0.00300)^2 - 4(1)(-2.0 \times 10^{-6})}}{2(1)} = 0.000562, \ -0.00356$$

The corresponding pH is 3.25:

$$\text{pH} = -\log[H_3O^+] = -\log x = 3.25$$

S16-18. Use the same method as in Exercise S16-15, this time solving the mass-action equation for various initial concentrations of formic acid (pK_a = 0.00018):

$$\text{HCOOH(aq)} + \text{H}_2\text{O}(\ell) \rightleftarrows \text{HCOO}^-\text{(aq)} + \text{H}_3\text{O}^+\text{(aq)}$$

$$[\text{HCOOH}]_0 - x \qquad\qquad x \qquad\qquad x$$

$$\frac{[\text{H}_3\text{O}^+][\text{HCOO}^-]}{[\text{HCOOH}]} = K_\text{a}$$

$$\frac{x^2}{[\text{HCOOH}]_0 - x} = 0.00018$$

From the value of x, we then obtain the pH and the anion-to-acid ratio:

$$\text{pH} = -\log[\text{H}_3\text{O}^+] = -\log x$$

$$\frac{[\text{HCOO}^-]}{[\text{HCOOH}]} = \frac{x}{[\text{HCOOH}]_0 - x}$$

Results are summarized below:

	$[\text{HCOOH}]_0$ (M)	pH	$\frac{[\text{HCOO}^-]}{[\text{HCOOH}]}$
(a)	1.000	1.88	0.014
(b)	0.500	2.03	0.019
(c)	0.100	2.38	0.043
(d)	0.010	2.90	0.14

S16-19. Apply the Henderson-Hasselbalch equation to a buffer solution composed of HCOO^- and HCOOH:

$$\text{pH} = \text{p}K_\text{a} + \log\frac{[\text{HCOO}^-]_0}{[\text{HCOOH}]_0}$$

(a) Making use of the following data,

$$\text{pH} = 3.80$$

$$\text{p}K_\text{a} = -\log K_\text{a} = -\log 0.00018 = 3.7447$$

$$[\text{HCOOH}]_0 = 1.000\ M$$

we solve for the concentration of HCOO^-:

$$\log\frac{[\mathrm{HCOO^-}]_0}{[\mathrm{HCOOH}]_0} = \mathrm{pH} - \mathrm{p}K_\mathrm{a}$$

$$\log[\mathrm{HCOO^-}]_0 = \mathrm{pH} - \mathrm{p}K_\mathrm{a} + \log[\mathrm{HCOOH}]_0 = 3.80 - 3.7447 + \log 1.000 = 0.05527$$

$$[\mathrm{HCOO^-}]_0 = 10^{0.05527} = 1.1357$$

Rounded to two significant figures, our result is 1.1 *M*.

(b) Add a slight excess of NaOH, symbolized by $\delta = 0.100$ *M*. The concentration of $HCOO^-$ goes up, and the concentration of HCOOH goes down. The pH increases modestly, from 3.80 to 3.88:

$$\mathrm{pH} = \mathrm{p}K_\mathrm{a} + \log\frac{[\mathrm{HCOO^-}]_0 + \delta}{[\mathrm{HCOOH}]_0 - \delta} = 3.7447 + \log\frac{1.1357 + 0.100}{1.000 - 0.100} = 3.88$$

(c) Addition of 0.100 *M* HCl produces the opposite effect. The pH decreases by 0.08 unit, from 3.80 to 3.72:

$$\mathrm{pH} = \mathrm{p}K_\mathrm{a} + \log\frac{[\mathrm{HCOO^-}]_0 - \delta}{[\mathrm{HCOOH}]_0 + \delta} = 3.7447 + \log\frac{1.1357 - 0.100}{1.000 + 0.100} = 3.72$$

S16-20. Another $HCOO^-$/HCOOH buffer solution.

(a) Convert grams into moles,

$$[\mathrm{HCOO^-}]_0 = \frac{6.80\ \mathrm{g\ NaHCOO}}{0.250\ \mathrm{L}} \times \frac{1\ \mathrm{mol\ NaHCOO}}{68.008\ \mathrm{g\ NaHCOO}} \times \frac{1\ \mathrm{mol\ HCOO^-}}{\mathrm{mol\ NaHCOO}} = 0.400\ M$$

$$[\mathrm{HCOOH}]_0 = \frac{5.75\ \mathrm{g\ HCOOH}}{0.250\ \mathrm{L}} \times \frac{1\ \mathrm{mol\ HCOOH}}{46.026\ \mathrm{g\ HCOOH}} = 0.500\ M$$

and substitute the resulting concentrations into the Henderson-Hasselbalch equation:

$$\mathrm{pH} = \mathrm{p}K_\mathrm{a} + \log\frac{[\mathrm{HCOO^-}]_0}{[\mathrm{HCOOH}]_0} = -\log 0.00018 + \log\frac{0.400}{0.500} = 3.65$$

(b) A solution prepared so that

$$\frac{[\mathrm{HCOO^-}]_0}{[\mathrm{HCOOH}]_0} = 0.569$$

will have a buffered pH of 3.50:

$$\log \frac{[\text{HCOO}^-]_0}{[\text{HCOOH}]_0} = \text{pH} - \text{p}K_\text{a} = 3.50 - 3.7447 = -0.2447$$

$$\frac{[\text{HCOO}^-]_0}{[\text{HCOOH}]_0} = 10^{-0.2447} = 0.569$$

To achieve the desired ratio, we need to add enough HCOOH to increase the total concentration to 0.703 M:

$$[\text{HCOOH}]_0 = \frac{[\text{HCOO}^-]_0}{10^{-0.2447}} = \frac{0.400}{0.569} = 0.703\ M$$

(c) Adding *strong* acid, HCl, we decrease the concentration of $HCOO^-$ and increase the concentration of HCOOH. A swing value of

$$\delta = 0.0736\ M\ \text{HCl}$$

will yield the same anion-to-acid ratio (and hence the same pH) calculated above in (b):

$$\frac{0.400 - \delta}{0.500 + \delta} = 0.569$$

$$0.400 - \delta = (0.500)(0.569) + 0.569\,\delta$$

$$\delta = 0.0736$$

S16-21. We consider the effect of volume on both the pH and the resiliency of a buffer solution.

(a) With equal concentrations of A^- and HA,

$$[\text{A}^-]_0 = \frac{1.000 \text{ mol}}{1.000 \text{ L}} = 1.000\ M$$

$$[\text{HA}]_0 = \frac{1.000 \text{ mol}}{1.000 \text{ L}} = 1.000\ M$$

the Henderson-Hasselbalch equation requires that pH = pK_a:

$$\text{pH} = \text{p}K_\text{a} + \log \frac{[\text{A}^-]_0}{[\text{HA}]_0} = -\log\left(1.00 \times 10^{-4}\right) + \log \frac{1.000}{1.000} = 4.000$$

(b) The pH barely decreases—from 4.000 to 3.999—upon addition of 0.001 mol HCl to the existing volume of 1.00 L:

$$\text{pH} = \text{p}K_a + \log\frac{[\text{A}^-]_0 - \delta}{[\text{HA}]_0 + \delta} = 4.000 + \log\frac{1.000 - 0.001}{1.000 + 0.001} = 3.999$$

(c) An increase in volume would not, by itself, change the pH of the buffer if the Henderson-Hasselbalch equation were still applicable:

$$\text{pH} = \text{p}K_a + \log\frac{[\text{A}^-]_0}{[\text{HA}]_0} = 4.000 + \log\left(\frac{\dfrac{1.000\text{ mol}}{100.000\text{ L}}}{\dfrac{1.000\text{ mol}}{100.000\text{ L}}}\right) = 4.000$$

At these low concentrations of 0.01000 *M*, however, we obtain a more accurate result by taking the approach set up below:

	$HA(aq)$	+	$H_2O(\ell)$	$\rightleftarrows$	$A^-(aq)$	+	$H_3O^+(aq)$
Initial conc.	0.01000				0.01000		0
Change	$-x$				x		x
Equil. conc.	$0.01000 - x$				$0.01000 + x$		x

The mass-action equation, a quadratic form, yields a pH of 4.009:

$$K_a = \frac{[\text{H}_3\text{O}^+][\text{A}^-]}{[\text{HA}]} = \frac{x(0.01000 + x)}{(0.01000 - x)} = 1.00 \times 10^{-4}$$

$$0 = x^2 + 0.0101x - \left(1.00 \times 10^{-6}\right)$$

$$x = \frac{-0.0101 \pm \sqrt{(0.0101)^2 - 4(1)\left(-1.00 \times 10^{-6}\right)}}{2(1)} = 9.806 \times 10^{-5},\ -1.020 \times 10^{-2}$$

$$\text{pH} = -\log[\text{H}_3\text{O}^+] = -\log x = -\log\left(9.806 \times 10^{-5}\right) = 4.009$$

(d) The buffer concentrations are diluted by a factor of 100,

$$[A^-]_0 = [HA]_0 = \frac{1.000 \text{ mol}}{100.000 \text{ L}} = 0.01000\ M$$

and the addition of 0.1 mol HCl (at constant volume) creates a stray H^+ concentration of 0.001 M in 100 L:

$$\delta = \frac{0.1 \text{ mol H}^+}{100.000 \text{ L}} = 0.001\ M$$

Given these altered conditions, the system then comes to a new equilibrium. As in part (c), we forgo the approximations of the Henderson-Hasselbalch equation in favor of a more exact treatment:

	$HA(aq)$	+ $H_2O(\ell)$ $\rightleftarrows$	$A^-(aq)$	+ $H_3O^+(aq)$
Initial conc.	$0.01000 + 0.001$		$0.01000 - 0.001$	0
Change	$-x$		x	x
Equil. conc.	$0.011 - x$		$0.009 + x$	x

Dilution of the buffer leaves it less resistant to the intrusion of strong acid. The change in pH—from 4.009 in part (c) to 3.923 in the present system—is larger than the decrease of 0.001 unit previously observed in part (b):

$$K_a = \frac{[H_3O^+][A^-]}{[HA]} = \frac{x(0.009 + x)}{(0.011 - x)} = 1.00 \times 10^{-4}$$

$$0 = x^2 + 0.0091x - \left(1.1 \times 10^{-6}\right)$$

$$x = \frac{-0.0091 \pm \sqrt{(0.0091)^2 - 4(1)\left(-1.1 \times 10^{-6}\right)}}{2(1)} = 1.193 \times 10^{-4},\ -9.219 \times 10^{-3}$$

$$\text{pH} = -\log[H_3O^+] = -\log x = -\log\left(1.193 \times 10^{-4}\right) = 3.923$$

S16-22. If an acid–base reaction is exothermic,

$$\text{HA(aq)} + \text{H}_2\text{O}(\ell) \rightleftarrows \text{A}^-\text{(aq)} + \text{H}_3\text{O}^+\text{(aq)} + \text{heat} \qquad \Delta H^\circ < 0$$

then the equilibrium constant decreases as the temperature increases:

$$K_a(T) = \exp\left(-\frac{\Delta G^\circ}{RT}\right) = \exp\left(-\frac{\Delta H^\circ}{RT}\right)\exp\left(\frac{\Delta S^\circ}{R}\right)$$

$$\frac{K_a(T_2)}{K_a(T_1)} = \exp\left[-\frac{\Delta H^\circ}{R}\left(\frac{1}{T_2} - \frac{1}{T_1}\right)\right] < 1 \quad \text{if} \quad \Delta H^\circ < 0 \text{ and } T_2 > T_1$$

Heat behaves as a product, shifting the process leftward—toward undissociated HA—at the higher temperature. The pH goes up as the concentration of hydronium ion goes down:

$$\text{pH} = -\log[\text{H}_3\text{O}^+]$$

S16-23. Since the reaction takes place entirely within a condensed phase,

$$\text{HA(aq)} + \text{H}_2\text{O}(\ell) \rightleftarrows \text{A}^-\text{(aq)} + \text{H}_3\text{O}^+\text{(aq)}$$

we assume that the position of its equilibrium is unaffected by pressure. The pH should remain the same at higher pressure.

S16-24. An increase in the volume of solvent will bring about a decrease in the initial concentration of HA and, ultimately, a decrease in the equilibrium concentration of H_3O^+. The pH will be higher in the larger volume.

S16-25. The residual free energy, ΔG, is *zero* for all reactions at equilibrium, no matter what the value of the equilibrium constant may be. The free energy of the products is equal to the free energy of the reactants. There is no drive to produce one species at the expense of another.

Chapter 17

Chemistry and Electricity

S17-1. Which of the following half-reactions will probably have the higher reduction potential?

$$XeO_3(aq) + 6H^+(aq) + 6e^- \rightarrow Xe(g) + 3H_2O(\ell)$$

$$Cu^{2+}(aq) + 2e^- \rightarrow Cu(s)$$

Use general chemical arguments—not a table of electrode potentials—to make your assessment.

S17-2. Which of the following half-reactions will probably have the higher reduction potential?

$$O_3(g) + 2H^+(aq) + 2e^- \rightarrow O_2(g) + H_2O(\ell)$$

$$K^+(aq) + e^- \rightarrow K(s)$$

Use general chemical arguments to make your assessment.

S17-3. Which of the following metals is oxidized more readily than zinc?

(a) Ag **(b)** Na **(c)** Al **(d)** Ni

S17-4. By custom, the reduction of H^+ to H_2 is assigned a standard potential of zero volts:

$$2H^+(aq, 1\ M) + 2e^- \rightarrow H_2(g, 1\ atm) \qquad \mathcal{E}^\circ_{red} = 0\ V\ \text{(conventional)}$$

Suppose, instead, that we use the reduction of Cu^{2+} to establish a new zero point:

$$Cu^{2+}(aq, 1\ M) + 2e^- \rightarrow Cu(s) \qquad \mathcal{E}^\circ_{red} = 0\ V\ \text{(rescaled)}$$

(a) What voltage would be obtained for the reduction of Zn^{2+} to Zn in the rescaled system? If necessary, consult the relevant table in Appendix C. **(b)** What voltage would be obtained for the zinc–copper reaction written below?

$$Zn(s) + Cu^{2+}(aq) \rightarrow Zn^{2+}(aq) + Cu(s)$$

(c) Is the rescaled voltage different from the conventional voltage?

S17-5. Species A, which exists in the forms A(s) and $A^{p+}(aq)$, has a standard reduction potential of 1.00 V. Species B, which exists as B(s) and $B^{q+}(aq)$, has a standard reduction potential of 2.00 V. **(a)** Write a balanced equation that depicts a spontaneous redox reaction between A and B. **(b)** Which species is the oxidizing agent? Which is the reducing agent? **(c)** Calculate the standard cell potential.

S17-6. Consider the two half-reactions below:

$$A(s) \rightarrow A^+(aq) + e^- \qquad \mathcal{E}^\circ_{ox} = 1.00\ V$$

$$3e^- + B^{3+}(aq) \rightarrow B(s) \qquad \mathcal{E}^\circ_{red} = -1.00\ V$$

(a) Calculate the standard cell potential for the process

$$3A(s) + B^{3+}(aq) \rightarrow 3A^+(aq) + B(s)$$

(b) Calculate ΔG° and K (the equilibrium constant). Is the reaction spontaneous under standard conditions?

S17-7. Consider the following half-reactions:

$$A(s) \rightarrow A^+(aq) + e^- \qquad \mathcal{E}^\circ_{ox} = 2.00\ V$$

$$4e^- + B^{4+}(aq) \rightarrow B(s) \qquad \mathcal{E}^\circ_{red} = 1.00\ V$$

(a) For which *separate* process, reduction or oxidation, is ΔG° more negative? Treat each half-reaction in isolation, as written. **(b)** Calculate $\mathcal{E}^\circ$ for the coupled reaction below:

$$4A(s) + B^{4+}(aq) \rightarrow 4A^+(aq) + B(s)$$

(c) Suppose that $[A^+] = 1.1\ M$ and $[B^{4+}] = 2.0\ M$. Is the resulting cell potential greater than, less than, or equal to $\mathcal{E}^\circ$?

S17-8. Consider an arbitrary reaction for which the standard cell potential is 2.00 V:

$$A^+(aq) + B(g) \rightarrow A(g) + B^+(aq) \qquad \mathcal{E}^\circ = 2.00 \text{ V}$$

The voltage is subsequently measured at 25°C over a range of partial pressures, with concentrations fixed at 1 *M*. Which of the following statements is always true?

(a) The cell potential is 2.00 V when $P_A = 1$ atm.
(b) The cell potential is 2.00 V when $P_B = 1$ atm.
(c) The cell potential is 2.00 V when $P_A = P_B$.
(d) The cell potential is 2.00 V only when $P_A = P_B = 1$ atm.

S17-9. Again, consider the reaction specified below:

$$A^+(aq) + B(g) \rightarrow A(g) + B^+(aq) \qquad \mathcal{E}^\circ = 2.00 \text{ V}$$

The voltage is measured at 25°C over a range of concentrations, with all partial pressures fixed at 1 atm. Which of the following statements is always true?

(a) The cell potential is 2.00 V when $[A^+] = 1$ *M*.
(b) The cell potential is 2.00 V when $[B^+] = 1$ *M*.
(c) The cell potential is 2.00 V when $[A^+] = [B^+]$.
(d) The cell potential is 2.00 V only when $[A^+] = [B^+] = 1$ *M*.

S17-10. For the last time:

$$A^+(aq) + B(g) \rightarrow A(g) + B^+(aq) \qquad \mathcal{E}^\circ = 2.00 \text{ V}$$

An experimenter measures a cell potential of 1.99 V under the conditions stated below:

$$P_A = P_B = 1 \text{ atm} \qquad [A^+] = 1\ M \qquad T = 298.15 \text{ K}$$

(a) Is the value of $[B^+]$ greater than, less than, or equal to 1 *M*? **(b)** Is the reaction spontaneous in the forward direction?

S17-11. Use the formation data in Table C-16 to calculate the standard oxidation potential for the following reaction:

$$Ag(s) + Cl^-(aq) \rightarrow AgCl(s) + e^-$$

S17-12. **(a)** Without consulting Table C-21, use the formation data in Table C-16 to calculate the standard cell potential for the reaction below:

$$Cu(s) + Cl_2(g) \rightarrow Cu^{2+}(aq) + 2Cl^-(aq)$$

(b) Recalculate the standard voltage, this time using the reduction potentials in Table C-21.

S17-13. The reduction potentials tabulated in Appendix C were measured under standard conditions: pressures of 1 atm, concentrations of 1 *M*, temperatures of 25°C. Recalculate the reduction potential for the half-reaction

$$Zn^{2+}(aq) + 2e^- \rightarrow Zn(s)$$

under the conditions stated below:

	$[Zn^{2+}]$ (*M*)	*P* (atm)	*T* (K)
(a)	0.100	1.00	298.15
(b)	10.000	1.00	298.15
(c)	1.000	0.10	298.15
(d)	1.000	10.00	298.15

S17-14. Recalculate the reduction potential of the hydrogen electrode under the conditions stated below:

	pH	P_{H_2} (atm)	*T* (K)
(a)	1.000	1.00	298.15
(b)	–1.000	1.00	298.15
(c)	0.000	0.10	298.15
(d)	0.000	10.00	298.15

S17-15. A certain Zn/Cu cell exhibits a potential of 2.00 V at 25°C. **(a)** Calculate the reaction quotient. **(b)** Is the system in equilibrium? If not, in which direction will the reaction proceed—toward products or toward reactants?

S17-16. A solution containing Cr^{3+} ions is subjected to electrolysis for 100.0 minutes. What current must be applied to deposit 3.65 grams of chromium metal?

S17-17. How much time is needed to produce 10.0 grams of each metal by application of an electrolytic current of 2.00 A?

(a) K from K^+ **(b)** Rb from Rb^+ **(c)** Mg from Mg^{2+} **(d)** Ti from Ti^{4+}

S17-18. How many moles of each metal will be obtained after 10.0 hours of electrolysis at 5.00 A?

(a) Ag from Ag^+ **(b)** K from K^+ **(c)** Ni from Ni^{2+} **(d)** Cu from Cu^{2+}

S17-19. Which of the following species, under normal electrochemical conditions, is able to act simultaneously as an oxidizing agent and a reducing agent?

(a) Fe^{2+} **(b)** Cu **(c)** Na^{+}

S17-20. **(a)** Write the following process as a sum of reduction and oxidation half-reactions:

$$2Cu^{+}(aq) \rightarrow Cu^{2+}(aq) + Cu(s)$$

(b) Which species is the oxidizing agent? Which is the reducing agent?

S17-21. The *disproportionation* reaction treated in the preceding exercise,

$$2Cu^{+}(aq) \rightarrow Cu^{2+}(aq) + Cu(s)$$

proceeds very rapidly in aqueous solution, usually coming to equilibrium in less than one second. **(a)** Use the formation data in Table C-16 to calculate the equilibrium constant at 25°C. **(b)** Which of two salts—CuCl or $CuCl_2$—would you expect to be insoluble in water? Explain your reasoning.

S17-22. Continue the analysis from the preceding exercise, and suppose (contrary to fact) that the reaction

$$2Cu^{+}(aq) \rightarrow Cu^{2+}(aq) + Cu(s)$$

proceeds very slowly, perhaps taking centuries to reach equilibrium. If so, would you still be prepared to comment on the relative solubilities of CuCl and $CuCl_2$?

S17-23. F_2 has a large, positive reduction potential. Li^{+} has a large, negative reduction potential. At equilibrium, state whether the residual free energy (ΔG) for the reduction of fluorine,

$$F_2(g) + 2e^{-} \rightleftarrows 2F^{-}(aq)$$

is greater than, less than, or equal to the residual free energy for the reduction of lithium:

$$Li^{+}(aq) + e^{-} \rightleftarrows Li(s)$$

Be sure to distinguish ΔG from ΔG°.

SOLUTIONS

S17-1. The reduction of XeO_3 to Xe—a noble gas, usually nonreactive—is expected to exhibit the higher, more favorable potential. Xenon is a thermodynamically more stable product than metallic copper.

S17-2. The reduction of O_3 to O_2 should show the higher potential. Potassium is a metal and hence more likely to be oxidized.

S17-3. Compare the oxidation potential of each metal with the oxidation potential of zinc:

$$Zn(s) \rightarrow Zn^{2+}(aq) + 2e^- \qquad \mathcal{E}^\circ_{ox}(Zn) = -\mathcal{E}^\circ_{red}(Zn^{2+}) = -(-0.76\text{ V}) = 0.76\text{ V}$$

Any oxidation for which $\mathcal{E}^\circ_{ox}$ is greater (more positive) than 0.76 V is favored over the Zn half-reaction. See Table C-21 in *Principles of Chemistry*.

HALF-REACTION	$\mathcal{E}^\circ_{ox}$ (V)	COMMENT
(a) $Ag(s) \rightarrow Ag^+(aq) + e^-$	–0.80	oxidizes less readily than Zn
(b) $Na(s) \rightarrow Na^+(aq) + e^-$	2.71	oxidizes more readily than Zn
(c) $Al(s) \rightarrow Al^{3+}(aq) + 3e^-$	1.66	oxidizes more readily than Zn
(d) $Ni(s) \rightarrow Ni^{2+}(aq) + 2e^-$	0.26	oxidizes less readily than Zn

S17-4. We demonstrate, by example, that the voltage scale is entirely arbitrary.

(a) The conventional voltages are reported in the following way:

$$Cu^{2+}(aq, 1\ M) + 2e^- \rightarrow Cu(s) \qquad \mathcal{E}^\circ_{red} = 0.34\text{ V}$$

$$Zn^{2+}(aq, 1\ M) + 2e^- \rightarrow Zn(s) \qquad \mathcal{E}^\circ_{red} = -0.76\text{ V}$$

The rescaled voltages would be shifted uniformly (down) by 0.34 V:

$$Cu^{2+}(aq, 1\ M) + 2e^- \rightarrow Cu(s) \qquad \mathcal{E}^\circ_{red} = 0.00\text{ V}$$

$$Zn^{2+}(aq, 1\ M) + 2e^- \rightarrow Zn(s) \qquad \mathcal{E}^\circ_{red} = -1.10\text{ V}$$

(b) The potential difference remains the same (1.10 V), regardless of how we define the zero point of voltage:

$$\begin{array}{ll} \text{Zn} \rightarrow \text{Zn}^{2+} + 2\text{e}^- & \mathcal{E}^\circ_{\text{ox}} = -(-1.10 \text{ V}) = 1.10 \text{ V} \\ 2\text{e}^- + \text{Cu}^{2+} \rightarrow \text{Cu} & \mathcal{E}^\circ_{\text{red}} = 0.00 \text{ V} \\ \hline \text{Zn(s)} + \text{Cu}^{2+}\text{(aq)} \rightarrow \text{Zn}^{2+}\text{(aq)} + \text{Cu(s)} & \mathcal{E}^\circ = \mathcal{E}^\circ_{\text{red}} + \mathcal{E}^\circ_{\text{ox}} = 1.10 \text{ V} \end{array}$$

(c) The rescaled voltage is no different from the conventional measure. All numerical relationships between electrode potentials are preserved intact.

S17-5. The species with the higher reduction potential (B^{q+}) undergoes reduction. Acting as the oxidizing agent, it takes electrons from its competitor.

(a) Electrons flow spontaneously from A to B^{q+}. To conserve charge, we multiply the oxidation half-reaction by q and the reduction half-reaction by p:

$$\begin{array}{ll} q[\text{A} \rightarrow \text{A}^{p+} + pe^-] & \mathcal{E}^\circ_{\text{ox}} = -(1.00 \text{ V}) \\ p[qe^- + \text{B}^{q+} \rightarrow \text{B}] & \mathcal{E}^\circ_{\text{red}} = 2.00 \text{ V} \\ \hline q\text{A(s)} + p\text{B}^{q+}\text{(aq)} \rightarrow q\text{A}^{p+}\text{(aq)} + p\text{B(s)} & \mathcal{E}^\circ = \mathcal{E}^\circ_{\text{red}} + \mathcal{E}^\circ_{\text{ox}} = 1.00 \text{ V} \end{array}$$

(b) Species B^{q+}, which undergoes reduction, is the oxidizing agent. Species A, which undergoes oxidation, is the reducing agent. See above.

(c) The cell delivers 1.00 volt:

$$\mathcal{E}^\circ = \mathcal{E}^\circ_{\text{red}} + \mathcal{E}^\circ_{\text{ox}} = 2.00 \text{ V} + (-1.00 \text{ V}) = 1.00 \text{ V}$$

Note that the oxidation potential for A is opposite in sign to the reduction potential for A^{p+}:

$$pe^- + \text{A}^{p+} \rightarrow \text{A} \qquad \mathcal{E}^\circ_{\text{red}} = 1.00 \text{ V}$$

$$\text{A} \rightarrow \text{A}^{p+} + pe^- \qquad \mathcal{E}^\circ_{\text{ox}} = -\mathcal{E}^\circ_{\text{red}} = -1.00 \text{ V}$$

S17-6. Here we recall the relationship between free energy, the cell potential, and the equilibrium constant:

$$\Delta G^\circ = -n\mathcal{F}\mathcal{E}^\circ = -RT \ln K$$

(a) Three electrons flow from reducing agent to oxidizing agent. The standard cell potential is exactly zero:

$$\begin{array}{ll} 3[\mathrm{A} \rightarrow \mathrm{A}^+ + \mathrm{e}^-] & \mathcal{E}^\circ_{\mathrm{ox}} = 1.00 \text{ V} \\ 3\mathrm{e}^- + \mathrm{B}^{3+} \rightarrow \mathrm{B} & \mathcal{E}^\circ_{\mathrm{red}} = -1.00 \text{ V} \\ \hline 3\mathrm{A(s)} + \mathrm{B}^{3+}\mathrm{(aq)} \rightarrow 3\mathrm{A}^+\mathrm{(aq)} + \mathrm{B(s)} & \mathcal{E}^\circ = \mathcal{E}^\circ_{\mathrm{red}} + \mathcal{E}^\circ_{\mathrm{ox}} = 0.00 \text{ V} \end{array}$$

(b) The standard change in free energy, directly proportional to $\mathcal{E}^\circ$, is zero as well:

$$\Delta G^\circ = -n\mathcal{F}\mathcal{E}^\circ = 0$$

Thus with concentrations equal to 1 M and with its equilibrium constant equal to 1,

$$K = \exp\left(-\frac{\Delta G^\circ}{RT}\right) = \exp 0 = 1$$

this system is already in equilibrium and shows no tendency to react at all. The process is neither spontaneous nor nonspontaneous, because there is no difference in free energy between products and reactants under standard conditions.

S17-7. Free energy is an extensive property, scaling in direct proportion to the moles of electrons transferred.

(a) The reduction half-reaction, by itself, exhibits the more negative value of ΔG°. A larger value of n compensates for a smaller value of $\mathcal{E}^\circ$:

	HALF-REACTION	n	$\mathcal{E}^\circ$ (V)	$\Delta G^\circ = -n\mathcal{F}\mathcal{E}^\circ$
Oxidation:	$\mathrm{A(s)} \rightarrow \mathrm{A}^+\mathrm{(aq)} + \mathrm{e}^-$	1	2.00	$-2.00\,\mathcal{F}$ V mol e^-
Reduction:	$4\mathrm{e}^- + \mathrm{B}^{4+}\mathrm{(aq)} \rightarrow \mathrm{B(s)}$	4	1.00	$-4.00\,\mathcal{F}$ V mol e^-

(b) Coupled together, the two half-reactions establish a net potential difference of 3.00 V:

$$\begin{array}{ll} 4[\mathrm{A} \rightarrow \mathrm{A}^+ + \mathrm{e}^-] & \mathcal{E}^\circ_{\mathrm{ox}} = 2.00 \text{ V} \\ 4\mathrm{e}^- + \mathrm{B}^{4+} \rightarrow \mathrm{B} & \mathcal{E}^\circ_{\mathrm{red}} = 1.00 \text{ V} \\ \hline 4\mathrm{A(s)} + \mathrm{B}^{4+}\mathrm{(aq)} \rightarrow 4\mathrm{A}^+\mathrm{(aq)} + \mathrm{B(s)} & \mathcal{E}^\circ = \mathcal{E}^\circ_{\mathrm{red}} + \mathcal{E}^\circ_{\mathrm{ox}} = 3.00 \text{ V} \end{array}$$

Voltage, measured as joules per coulomb, is an intensive property: $\mathcal{E}^\circ_{\mathrm{red}}$ and $\mathcal{E}^\circ_{\mathrm{ox}}$ add together directly, regardless of the total number of electrons transferred.

(c) Insert the reaction quotient

$$Q = \frac{[A^+]^4}{[B^{4+}]} = \frac{(1.1)^4}{2.0} = 0.732$$

into the Nernst equation, and evaluate the instantaneous cell potential:

$$\mathcal{E} = \mathcal{E}^\circ - \frac{RT}{n\mathcal{F}} \ln Q = \mathcal{E}^\circ - \frac{RT}{n\mathcal{F}} \ln 0.732 = \mathcal{E}^\circ + \frac{0.312\, RT}{n\mathcal{F}} > \mathcal{E}^\circ$$

S17-8. Choice (c) is correct. If $P_A = P_B$, then $Q = 1$ and $\mathcal{E} = \mathcal{E}^\circ$:

$$Q = \frac{P_A[B^+]}{[A^+]P_B} = \frac{P_A(1)}{(1)P_B} = \frac{P_A}{P_B} = 1 \qquad \text{if} \qquad P_A = P_B$$

$$\mathcal{E} = \mathcal{E}^\circ - \frac{RT}{n\mathcal{F}} \ln Q = \mathcal{E}^\circ \qquad \text{if} \qquad Q = 1$$

Note that choice (d) is too restrictive. The condition $P_A = P_B = 1$ atm is sufficient but not necessary.

S17-9. Choice (c) is correct, by the same reasoning used in the preceding exercise.

S17-10. We continue to apply the Nernst equation

$$\mathcal{E} = \mathcal{E}^\circ - \frac{RT}{n\mathcal{F}} \ln Q = \mathcal{E}^\circ - \frac{0.05916\text{ V}}{n} \log Q \qquad (T = 298.15\text{ K})$$

to the hypothetical reaction treated in the last two exercises:

$$A^+(aq) + B(g) \rightarrow A(g) + B^+(aq) \qquad \mathcal{E}^\circ = 2.00\text{ V}$$

(a) Given $\mathcal{E}^\circ$ and $\mathcal{E}$, we determine that the reaction quotient is greater than 1:

$$\log Q = \frac{n}{0.05916\text{ V}}\left(\mathcal{E}^\circ - \mathcal{E}\right) = \frac{1}{0.05916\text{ V}}\left(2.00\text{ V} - 1.99\text{ V}\right) = 0.169 > 0$$

$$Q > 1$$

Doing so, we also find that the concentration of B^+ is greater than 1:

$$Q = \frac{P_A[B^+]}{[A^+]P_B}$$

$$[B^+] = \frac{[A^+]P_B}{P_A} Q = \frac{(1)(1)}{(1)} Q > 1$$

(b) Despite the small difference between $\mathcal{E}$ and $\mathcal{E}^\circ$, the reaction is still spontaneous in the forward direction. The instantaneous cell potential is positive.

S17-11. First, calculate the free energy of reaction by summing over the Gibbs formation energies:

$$Ag(s) + Cl^-(aq) \rightarrow AgCl(s) + e^-$$

$$\begin{aligned}\Delta G^\circ &= \Delta G_f^\circ[AgCl(s)] - \Delta G_f^\circ[Ag(s)] - \Delta G_f^\circ[Cl^-(aq)] \\ &= \left(-\frac{109.8 \text{ kJ}}{\text{mol}} \times 1 \text{ mol}\right) - \left(\frac{0 \text{ kJ}}{\text{mol}} \times 1 \text{ mol}\right) - \left(-\frac{131.2 \text{ kJ}}{\text{mol}} \times 1 \text{ mol}\right) = 21.4 \text{ kJ} \\ &= 2.14 \times 10^4 \text{ J}\end{aligned}$$

Second, calculate the corresponding voltage:

$$\mathcal{E}^\circ = -\frac{\Delta G^\circ}{n\mathcal{F}} = -\frac{2.14 \times 10^4 \text{ J}}{1 \text{ mol e}^- \times \dfrac{96{,}485 \text{ C}}{\text{mol e}^-}} = -0.222 \text{ J C}^{-1} = -0.222 \text{ V}$$

S17-12. Similar to the preceding exercise.

(a) Calculate ΔG° by summing over the free energies of formation:

$$Cu(s) + Cl_2(g) \rightarrow Cu^{2+}(aq) + 2Cl^-(aq)$$

$$\begin{aligned}\Delta G^\circ &= \Delta G_f^\circ[Cu^{2+}(aq)] + 2\,\Delta G_f^\circ[Cl^-(aq)] - \Delta G_f^\circ[Cu(s)] - \Delta G_f^\circ[Cl_2(g)] \\ &= \left(\frac{65.5 \text{ kJ}}{\text{mol}} \times 1 \text{ mol}\right) + \left(-\frac{131.2 \text{ kJ}}{\text{mol}} \times 2 \text{ mol}\right) - \left(\frac{0 \text{ kJ}}{\text{mol}} \times 1 \text{ mol}\right) - \left(\frac{0 \text{ kJ}}{\text{mol}} \times 1 \text{ mol}\right) \\ &= -196.9 \text{ kJ}\end{aligned}$$

Two electrons are transferred between Cu and Cl_2:

$$\mathcal{E}^\circ = -\frac{\Delta G^\circ}{n\mathcal{F}} = -\frac{(-1.969 \times 10^5 \text{ J})}{2 \text{ mol e}^- \times \frac{96{,}485 \text{ C}}{\text{mol e}^-}} = 1.020 \text{ J C}^{-1} = 1.020 \text{ V}$$

(b) The process splits into the following two half-reactions:

$$\begin{array}{ll} \text{Cu} \rightarrow \text{Cu}^{2+} + 2\text{e}^- & \mathcal{E}^\circ_{\text{ox}} = -0.3419 \text{ V} \\ 2\text{e}^- + \text{Cl}_2 \rightarrow 2\text{Cl}^- & \mathcal{E}^\circ_{\text{red}} = 1.3583 \text{ V} \\ \hline \text{Cu(s)} + \text{Cl}_2\text{(g)} \rightarrow \text{Cu}^{2+}\text{(aq)} + 2\text{Cl}^-\text{(aq)} & \mathcal{E}^\circ = \mathcal{E}^\circ_{\text{red}} + \mathcal{E}^\circ_{\text{ox}} = 1.0164 \text{ V} \end{array}$$

S17-13. Gaining two electrons, the Zn^{2+} ion is reduced to zinc metal:

$$\text{Zn}^{2+}\text{(aq)} + 2\text{e}^- \rightarrow \text{Zn(s)}$$

$$\mathcal{E}^\circ = -0.7618 \text{ V} \qquad n = 2 \qquad Q = \frac{1}{[\text{Zn}^{2+}]}$$

We insert these values into the Nernst equation and thus determine the nonstandard cell potentials at 25°C (298.15 K):

$$\mathcal{E} = \mathcal{E}^\circ - \frac{RT}{n\mathcal{F}} \ln Q = -0.7618 \text{ V} - \frac{0.05916 \text{ V}}{2} \log \frac{1}{[\text{Zn}^{2+}]} \qquad (T = 298.15 \text{ K})$$

Note that the voltage is independent of pressure in this reaction.

	$[Zn^{2+}]$ (M)	P (atm)	T (K)	Q	log Q	$\mathcal{E}$ (V)
(a)	0.100	1.00	298.15	10.0	1.000	−0.7914
(b)	10.000	1.00	298.15	0.10000	−1.00000	−0.7322
(c)	1.000	0.10	298.15	1.000	0.0000	−0.7618
(d)	1.000	10.00	298.15	1.000	0.0000	−0.7618

S17-14. Similar to the preceding exercise. Here we evaluate the Nernst equation

$$\mathcal{E} = \mathcal{E}^\circ - \frac{RT}{n\mathcal{F}} \ln Q = \mathcal{E}^\circ - \frac{0.05916 \text{ V}}{n} \log Q \qquad (T = 298.15 \text{ K})$$

for reduction of the hydrogen cation, H^+, to molecular hydrogen:

$$2H^+(aq) + 2e^- \rightarrow H_2(g)$$

$$\mathcal{E}^\circ = 0.0000 \text{ V} \qquad n = 2 \qquad Q = \frac{P_{H_2}}{[H^+]^2} = \frac{P_{H_2}}{\left(10^{-pH}\right)^2}$$

Results are summarized below:

	pH	P_{H_2} (atm)	T (K)	Q	log Q	$\mathcal{E}$ (V)
(a)	1.000	1.00	298.15	1.00×10^2	2.000	–0.0592
(b)	–1.000	1.00	298.15	1.00×10^{-2}	–2.000	0.0592
(c)	0.000	0.10	298.15	1.0×10^{-1}	–1.00	0.0296
(d)	0.000	10.00	298.15	1.00×10^1	1.000	–0.0296

S17-15. Two electrons flow between Zn and Cu^{2+} in our prototypical redox reaction:

$$Zn(s) + Cu^{2+}(aq) \rightleftarrows Zn^{2+}(aq) + Cu(s) \qquad (\mathcal{E}^\circ = 1.1037 \text{ V}, \ n = 2; \ K > 1)$$

(a) Rearrange the Nernst equation to solve for Q:

$$\mathcal{E} = \mathcal{E}^\circ - \frac{RT}{n\mathcal{F}} \ln Q = \mathcal{E}^\circ - \frac{0.05916 \text{ V}}{n} \log Q \qquad (\text{at } 25^\circ \text{C})$$

$$\log Q = \frac{n}{0.05916 \text{ V}}\left(\mathcal{E}^\circ - \mathcal{E}\right) = \frac{2}{0.05916 \text{ V}}\left(1.1037 \text{ V} - 2.00 \text{ V}\right) = -30.3$$

$$Q = 10^{\log Q} = 5 \times 10^{-31}$$

(b) With $\mathcal{E} > 0$, the system is not yet in equilibrium. The reaction will proceed to the right—toward products—until Q (currently less than 1) becomes equal to K.

S17-16. Divide the total charge by the elapsed time to obtain the current:

$$\text{Current} = \frac{\text{charge}}{\text{time}} = \frac{3.65 \text{ g Cr} \times \dfrac{1 \text{ mol Cr}}{51.9961 \text{ g Cr}} \times \dfrac{3 \text{ mol e}^-}{\text{mol Cr}} \times \dfrac{96{,}485 \text{ C}}{\text{mol e}^-}}{100.0 \text{ min} \times \dfrac{60 \text{ s}}{\text{min}}} \times \frac{1 \text{ A}}{\text{C s}^{-1}} = 3.39 \text{ A}$$

S17-17. From the mass of product, we calculate the charge needed to reduce each ion M^{n+} to the native metal M. From the current, we then calculate the elapsed time:

$$\text{Time} = \frac{\text{charge}}{\text{current}}$$

(a) The potassium ion, K^+, gains one electron during reduction to its metallic form:

$$\text{Time} = \frac{10.0 \text{ g K} \times \dfrac{1 \text{ mol K}}{39.0983 \text{ g K}} \times \dfrac{1 \text{ mol e}^-}{\text{mol K}} \times \dfrac{96{,}485 \text{ C}}{\text{mol e}^-}}{2.00 \text{ C s}^{-1}} = 1.23 \times 10^4 \text{ s} \quad (3.43 \text{ h})$$

(b) The rubidium ion, similar to the potassium ion, gains one electron during the reduction,

$$Rb^+(aq) + e^- \rightarrow Rb(s)$$

but the requisite time is not the same. The difference arises from the difference in molar mass:

$$\text{Time} = \frac{10.0 \text{ g Rb} \times \dfrac{1 \text{ mol Rb}}{85.4678 \text{ g Rb}} \times \dfrac{1 \text{ mol e}^-}{\text{mol Rb}} \times \dfrac{96{,}485 \text{ C}}{\text{mol e}^-}}{2.00 \text{ C s}^{-1}} = 5.64 \times 10^3 \text{ s} \quad (1.57 \text{ h})$$

(c) Two electrons are transferred to each Mg^{2+} ion:

$$\text{Time} = \frac{10.0 \text{ g Mg} \times \dfrac{1 \text{ mol Mg}}{24.305 \text{ g Mg}} \times \dfrac{2 \text{ mol e}^-}{\text{mol Mg}} \times \dfrac{96{,}485 \text{ C}}{\text{mol e}^-}}{2.00 \text{ C s}^{-1}} = 3.97 \times 10^4 \text{ s} \quad (11.0 \text{ h})$$

(d) One more example, this time with $n = 4$ (Ti from Ti^{4+}):

$$\text{Time} = \frac{10.0 \text{ g Ti} \times \dfrac{1 \text{ mol Ti}}{47.867 \text{ g Ti}} \times \dfrac{4 \text{ mol e}^-}{\text{mol Ti}} \times \dfrac{96{,}485 \text{ C}}{\text{mol e}^-}}{2.00 \text{ C s}^{-1}} = 4.03 \times 10^4 \text{ s} \quad (11.2 \text{ h})$$

S17-18. Variation on a theme. Given the current and time, we calculate the molar amount produced from each species M^{n+}.

(a) $$\frac{5.00 \text{ C}}{\text{s}} \times \frac{3600 \text{ s}}{\text{h}} \times 10.0 \text{ h} \times \frac{1 \text{ mol e}^-}{96{,}485 \text{ C}} \times \frac{1 \text{ mol Ag}}{\text{mol e}^-} = 1.87 \text{ mol Ag}$$

(b) $$\frac{5.00 \text{ C}}{\text{s}} \times \frac{3600 \text{ s}}{\text{h}} \times 10.0 \text{ h} \times \frac{1 \text{ mol e}^-}{96{,}485 \text{ C}} \times \frac{1 \text{ mol K}}{\text{mol e}^-} = 1.87 \text{ mol K}$$

(c) $$\frac{5.00\text{ C}}{\text{s}} \times \frac{3600\text{ s}}{\text{h}} \times 10.0\text{ h} \times \frac{1\text{ mol e}^-}{96{,}485\text{ C}} \times \frac{1\text{ mol Ni}}{2\text{ mol e}^-} = 0.933\text{ mol Ni}$$

(d) $$\frac{5.00\text{ C}}{\text{s}} \times \frac{3600\text{ s}}{\text{h}} \times 10.0\text{ h} \times \frac{1\text{ mol e}^-}{96{,}485\text{ C}} \times \frac{1\text{ mol Cu}}{2\text{ mol e}^-} = 0.933\text{ mol Cu}$$

S17-19. An oxidizing agent gains electrons; a reducing agent loses them. Under normal conditions, a species existing in its most positive oxidation state can only *gain* electrons.

(a) Fe^{2+} is reduced to Fe and oxidized to Fe^{3+}. The one species, iron(II), acts as both oxidizing agent and reducing agent:

$$\text{Oxidizing agent:}\quad 2e^- + Fe^{2+}(aq) \rightarrow Fe(s) \qquad \mathcal{E}^\circ_{\text{red}} = -0.447\text{ V}$$

$$\text{Reducing agent:}\quad Fe^{2+}(aq) \rightarrow Fe^{3+}(aq) + e^- \qquad \mathcal{E}^\circ_{\text{ox}} = -0.770\text{ V}$$

(b) Metallic copper is oxidized to a cation (Cu^+, Cu^{2+}, Cu^{3+}) but is not normally reduced to an anion. Cu functions only as a reducing agent.

(c) The sodium ion, Na^+, exists in a maximum oxidation state; it can be reduced to Na but not oxidized to Na^{2+} under normal conditions. Na^+ serves only as an oxidizing agent.

S17-20. In a *disproportionation* reaction, illustrated in this exercise, the same species acts simultaneously as oxidizing agent and reducing agent.

(a) In one half-reaction, the Cu^+ ion loses an electron and becomes Cu^{2+}. In the other half-reaction, it gains an electron and becomes Cu:

$$\text{Oxidation:}\quad Cu^+ \rightarrow Cu^{2+} + e^-$$

$$\text{Reduction:}\quad e^- + Cu^+ \rightarrow Cu$$

$$2Cu^+(aq) \rightarrow Cu^{2+}(aq) + Cu(s)$$

(b) Cu^+ is both oxidizing agent and reducing agent, as demonstrated above.

S17-21. We continue with the disproportionation of the copper(I) ion:

$$2Cu^+(aq) \rightarrow Cu^{2+}(aq) + Cu(s)$$

(a) The standard change in free energy is large and negative:

$$\Delta G^\circ = \Delta G_f^\circ\left[\text{Cu}^{2+}(\text{aq})\right] + \Delta G_f^\circ\left[\text{Cu(s)}\right] - 2\,\Delta G_f^\circ\left[\text{Cu}^{+}(\text{aq})\right]$$

$$= \left(\frac{65.5 \text{ kJ}}{\text{mol}} \times 1 \text{ mol}\right) + \left(\frac{0 \text{ kJ}}{\text{mol}} \times 1 \text{ mol}\right) - \left(\frac{50.0 \text{ kJ}}{\text{mol}} \times 2 \text{ mol}\right) = -34.5 \text{ kJ}$$

$$= -3.45 \times 10^4 \text{ J} \quad \text{(per mole of Cu)}$$

As a result, the equilibrium constant is substantially greater than 1:

$$K = \exp\left(-\frac{\Delta G^\circ}{RT}\right) = \exp\left[-\frac{\left(-3.45 \times 10^4 \text{ J mol}^{-1}\right)}{\left(8.3145 \text{ J mol}^{-1}\text{ K}^{-1}\right)\left(298.15 \text{ K}\right)}\right] = 1.11 \times 10^6$$

(b) We expect CuCl to be insoluble in water, because the Cu^+ ion is severely disadvantaged in aqueous solution. Cu^+ is rapidly converted to Cu^{2+} and Cu by disproportionation. The equilibrium, attained quickly, lies well to the right.

S17-22. From a thermodynamic standpoint, the Cu^+ ion is unstable relative to Cu^{2+} and Cu. If equilibrium can be established, then only a small amount of Cu^+ will coexist with Cu^{2+}.

As a practical matter, though, an exceedingly slow mechanism (such as the one posited here) can delay and effectively block the onset of equilibrium. Under the circumstances suggested, we would be able to observe Cu^+ in solution—and, with the system not in equilibrium, we would then have no thermodynamic basis for predicting the relative amounts of Cu^+ and Cu^{2+}.

S17-23. The residual free energy, ΔG, is *zero* for all reactions at equilibrium, no matter what the value of the equilibrium constant may be. The free energy of the products is equal to the free energy of the reactants. There is no drive to produce one species at the expense of another.

Chapter 18

Kinetics—The Course of Chemical Reactions

S18-1. Explain the meaning of the following terms, pointing out how each is different: *rate*, *rate law*, *rate constant*.

S18-2. Is it possible for the reaction

$$A + B \rightarrow C + D + E$$

to be described by one or more of the rate laws listed below? If so, which ones?

$$\text{Rate} = k[A]$$

$$\text{Rate} = k[A][B]$$

$$\text{Rate} = k[A]^2[B]$$

$$\text{Rate} = \frac{k_1[A]^{3/2}[C]}{[D]^{5/4} - k_2[E]} + k_3[A]$$

S18-3. **(a)** Assign correct units to the rate constant in the following expression:

$$\text{Rate} = k\frac{[A][B]}{[C]^2[D]^2}$$

(b) What is the overall order of reaction?

S18-4. Something new: If the kinetics of a reaction

$$\text{A} \rightarrow \text{products}$$

are ***zeroth*** order, the rate law is given by

$$\text{Rate} = -\frac{\Delta[\text{A}]}{\Delta t} = k$$

(a) What are the proper units for k? **(b)** If the initial concentration of A is $[\text{A}]_0$, convince yourself that the concentration at time t is given by

$$[\text{A}]_t = -kt + [\text{A}]_0$$

(c) Write an expression for the half-life. **(d)** At what time will $[\text{A}]_t$ fall to the value 0?

S18-5. Assume that the reaction

$$\text{A} \rightarrow \text{products}$$

follows zeroth-order kinetics, as described in the preceding exercise. Concentrations at $t = 10$ s and $t = 100$ s are found to be 1.00 M and 0.75 M, respectively. **(a)** Calculate k. **(b)** Calculate $[\text{A}]_0$. **(c)** Calculate $t_{1/2}$.

S18-6. Suppose that the initial rate of a hypothetical reaction

$$\text{A} + \text{B} + \text{C} \rightarrow \text{D} + \text{E}$$

is measured as follows:

TRIAL	[A]	[B]	[C]	INITIAL RATE ($M\,\text{s}^{-1}$)
1	0.10	0.10	0.10	0.484
2	0.20	0.10	0.10	0.483
3	0.10	0.20	0.10	0.967
4	0.10	0.10	0.20	1.936

All concentrations are in moles per liter ($1\ M = 1\ \text{mol L}^{-1}$). **(a)** Determine the kinetic order with respect to each reactant. **(b)** What is the overall rate law and order of reaction?

S18-7. Suppose that the reaction

$$2A \rightarrow \text{products}$$

obeys the following initial rate law:

$$\text{Rate} = k[A]^2$$

The initial concentration of A is 2.00 *M*, and the first half-life is 155.1 s. **(a)** Calculate the concentration at $t = 30.0$ s. **(b)** Is it possible for the reaction to be elementary? Can you tell?

S18-8. Say that the initial rate law for a reaction

$$A + B \rightarrow C$$

is second order in [A]:

$$\text{Rate} = k[A]^2$$

(a) Must this same rate law be maintained at all times? What additional variable or variables might become important as the process nears equilibrium? **(b)** Is it possible for the reaction to be elementary?

S18-9. Reaction 1 has an activation energy equal to 100 kJ mol^{-1}. For reaction 2, the activation energy is 10 kJ mol^{-1}. Further information, such as concerning collision frequency and efficiency, is not available. **(a)** Do you have sufficient data to predict which of the two reactions will have the larger rate constant at 25°C? If yes, do so. If no, specify what additional information is needed. **(b)** Do you have sufficient data to predict whether each reaction will be endothermic or exothermic? If yes, do so. If no, specify what else is needed. **(c)** Do you have sufficient data to predict whether the equilibrium constant for each reaction will be greater than 1, less than 1, or equal to 1? If yes, do so. If no, specify what more is needed.

S18-10. The rate constant $k(T)$ for a certain reaction obeys the Arrhenius law. Assume further that the pre-exponential factor is independent of temperature. **(a)** Does $k(T)$ increase, decrease, or remain the same as the temperature is increased? **(b)** Show that the relationship

$$\frac{1}{T_2} = \frac{1}{T_1} - \frac{R \ln p}{E_a}$$

holds when

$$\frac{k(T_2)}{k(T_1)} = p$$

S18-11. Use the results from Exercise S18-10 to answer the following questions. **(a)** Suppose that $E_a = 10$ kJ mol^{-1}. At what temperature will the value of k be tripled relative to its value at $T_1 = 300$ K? **(b)** Suppose, instead, that $E_a = 100$ kJ mol^{-1}. At what temperature will the value of k be tripled relative to its value at $T_1 = 300$ K?

S18-12. The rate constant for a certain first-order reaction is 0.500 s^{-1} at 400 K. **(a)** Calculate the half-life. **(b)** If the initial concentration is 0.750 *M*, what concentration remains after 3.25 s? **(c)** Do you expect the half-life to increase, decrease, or remain the same as the temperature is increased? **(d)** Do you expect the half-life to increase, decrease, or remain the same as the initial concentration is increased?

S18-13. Assume that the rate constant for some first-order reaction obeys the Arrhenius law. **(a)** Calculate the half-life at 500 K, given the following values for the pre-exponential factor (A) and the activation energy (E_a):

$$A = 1.00 \times 10^{12}\ \text{s}^{-1} \qquad\qquad E_a = 50.0\ \text{kJ mol}^{-1}$$

(b) Calculate the half-life at 600 K.

S18-14. Carried out in the gas phase, the reaction

$$\text{A} + \text{B} \rightarrow \text{C} + \text{D}$$

was found to obey the rate law

$$\text{Rate} = k[\text{A}][\text{B}]$$

with $k = 10^3\ M^{-1}\ \text{s}^{-1}$. **(a)** A priori—given no other information—would you expect the rate law to have the same algebraic form and order if the reaction were carried out in solution? **(b)** Would you expect the numerical value of k to be the same in solution as it is in the gas phase? **(c)** Would you expect the mechanism to be the same in solution as it is in the gas phase?

S18-15. Suppose that the equilibrium constant for some elementary reaction

$$\text{A} \rightleftarrows \text{B}$$

is exceedingly large—as large as, say, $K = 10^{50}$. At equilibrium, is the rate of the reverse reaction effectively zero? Does the reverse reaction *never* occur?

S18-16. Here are the forward and reverse rate constants (k_+ and k_-) for two elementary reactions:

REACTION	k_+ (s^{-1})	k_- (s^{-1})
$A \rightleftarrows B$	10	20
$C \rightleftarrows D$	100	50

(a) Calculate the equilibrium constant for each reaction. **(b)** Calculate the overall rate of each reaction at equilibrium.

S18-17. Assume that reaction 1 is elementary,

$$1. \quad A \rightleftarrows B$$

whereas reaction 2 is not:

$$2. \quad 3A + 2B + C \rightleftarrows 4D + E + F$$

The equilibrium constants are $K_1 = 100$ and $K_2 = 0.01$, respectively, for the two processes. **(a)** At equilibrium, is the overall rate of reaction 2 greater than, less than, or equal to the rate of reaction 1? **(b)** Can you determine a numerical value for the rate constant of reaction 1 in the forward direction? If yes, do so. If no, explain why. **(c)** Can you determine a numerical value for the rate constant of reaction 2 in the forward direction? If yes, do so. If no, explain why.

S18-18. Listed below are the free energies of reaction and activation for two hypothetical transformations, each process beginning with the same set of reactants:

REACTION	$\Delta G°$ (kJ mol^{-1})	$\Delta G^‡$ (kJ mol^{-1})
$R \rightleftarrows P_1$	−100	100
$R \rightleftarrows P_2$	−10	10

(a) Which reaction reaches equilibrium faster? **(b)** Which reaction yields the greater proportion of products at equilibrium?

S18-19. Answer the following questions, taking into account the thermodynamic data presented in Exercise S18-18. **(a)** Imagine that R, P_1, and P_2 come to a three-way equilibrium:

$$P_1 \rightleftarrows R \rightleftarrows P_2$$

Which product—P_1 or P_2—is found in the greater proportion at equilibrium? **(b)** Which product will predominate in the early stages of the three-way reaction? Is the process initially controlled by thermodynamics or kinetics?

S18-20. Again, consider the three species described in the preceding two exercises:

REACTION	$\Delta G°$ (kJ mol^{-1})	$\Delta G^{\ddagger}$ (kJ mol^{-1})
$R \rightleftarrows P_1$	-100	100
$R \rightleftarrows P_2$	-10	10

This time, however, imagine that the product P_2 (a gas) disappears from the system as soon as it is produced. **(a)** Is the process ever able to reach the system-wide equilibrium depicted below?

$$P_1 \rightleftarrows R \rightleftarrows P_2$$

(b) Do you expect the eventual amount of P_2 produced to be greater than, less than, or the same as the amount of P_1? Is the process controlled by thermodynamics or kinetics?

S18-21. An experimenter studying the hypothetical process

$$A + 2B + C \rightarrow D$$

finds that the average rate—as measured by the disappearance of A—is 0.001 M s^{-1} during the first 1.0 s of reaction:

$$-\frac{\Delta[A]}{\Delta t} = 0.001\ M\ s^{-1}$$

During this same interval, however, the product D fails to appear. The average initial rate—as measured by the *appearance* of D—is therefore zero:

$$\frac{\Delta[D]}{\Delta t} = 0$$

Suggest a reason for the inequality between these two rates.

S18-22. What's wrong with the picture below?

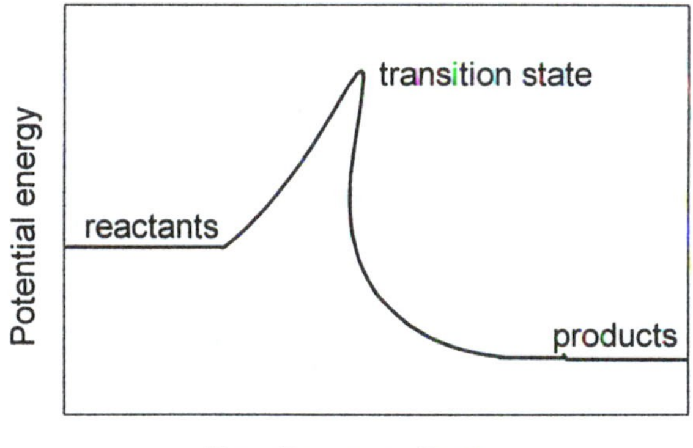

SOLUTIONS

S18-1. From the Glossary:

Rate. The change in a quantity per unit time. Examples: reaction rate = concentration/time; speed = distance/time.

Rate law. A mathematical relationship between the instantaneous rate of a reaction and the concentrations of the species involved. The expression is often (but not always) in the form

$$\text{Rate} = k[\text{A}]^n[\text{B}]^m\ldots$$

for reactants A, B, … .

Rate constant. The proportionality constant that appears in a rate law, connecting the rate of reaction with the concentrations of the relevant species. Symbol: k.

S18-2. All of the proposed rate laws are possible. Until we ascertain the rate law experimentally, we have no reason (in principle) to exclude any of the rate equations shown.

S18-3. We consider an arbitrary rate law that takes the following form:

$$\text{Rate} = k\frac{[\text{A}][\text{B}]}{[\text{C}]^2[\text{D}]^2}$$

(a) The units of k must be consistent with an overall rate expressed in $M\ \text{s}^{-1}$:

$$k = \text{rate} \times \frac{[\text{C}]^2[\text{D}]^2}{[\text{A}][\text{B}]} \sim M\ \text{s}^{-1} \times \frac{(M)^2(M)^2}{(M)(M)} \sim M^3\ \text{s}^{-1}$$

(b) The overall order of reaction is equal to the sum of the four exponents:

$$\text{Rate} = k[\text{A}]^1[\text{B}]^1[\text{C}]^{-2}[\text{D}]^{-2}$$

$$\text{Order} = 1 + 1 - 2 - 2 = -2$$

S18-4. Our presumed zeroth-order reaction

$$\text{A} \rightarrow \text{products}$$

proceeds at a constant rate, unaffected by the concentration of A:

$$\text{Rate} = -\frac{\Delta[\text{A}]}{\Delta t} = k$$

(a) The rate constant k has the same units as the rate itself: $M\ \text{s}^{-1}$.

(b) Consider the disappearance of A:

$$\frac{\Delta[\text{A}]}{\Delta t} = -k$$

Since the rate is constant at all times, we know that [A] decays along a straight line with slope equal to $-k$. The intercept on the [A] axis is simply the concentration at $t = 0$, symbolized as $[\text{A}]_0$. Knowing slope and intercept, we can then cast the linear equation in the general form $y = mx + b$:

$$[\text{A}]_t = -kt + [\text{A}]_0$$

(c) Solve for the time at which $[A]_t = \frac{1}{2}[A]_0$:

$$[A]_t = -kt + [A]_0$$

$$[A]_{t\,1/2} = \frac{[A]_0}{2} = -kt_{1/2} + [A]_0$$

$$kt_{1/2} = \frac{[A]_0}{2}$$

$$t_{1/2} = \frac{[A]_0}{2k}$$

(d) Solve for the time at which $[A]_t = 0$:

$$[A]_{t_\infty} = -kt_\infty + [A]_0 = 0$$

$$t_\infty = \frac{[A]_0}{k}$$

S18-5. We carry over the zeroth-order solution from Exercise S18-4:

$$\frac{\Delta[A]}{\Delta t} = -k$$

$$[A]_t = -kt + [A]_0$$

(a) Insert the data into the concentration profile, and subtract one equation from the other:

$$\begin{array}{rl} & 1.00\ M = -k(10\ \text{s}) + [A]_0 \\ - & \\ & 0.75\ M = -k(100\ \text{s}) + [A]_0 \\ \hline & 0.25\ M = k(90\ \text{s}) \end{array}$$

Solve for k:

$$k = \frac{0.25\ M}{90\ \text{s}} = 0.0028\ M\ \text{s}^{-1}$$

(b) First, add the two equations:

$$\begin{array}{rl} & 1.00\ M = -k(10\ \text{s}) + [A]_0 \\ + & \\ & 0.75\ M = -k(100\ \text{s}) + [A]_0 \\ \hline & 1.75\ M = -k(110\ \text{s}) + 2[A]_0 \end{array}$$

Then substitute $k = (0.25/90)\ M\ s^{-1}$ and solve for $[A]_0$:

$$[A]_0 = \frac{1}{2}\left[\left(\frac{0.25\ M}{90\ s}\right)(110\ s) + 1.75\ M\right] = 1.03\ M$$

(c) Use $[A]_0$ and k to calculate $t_{1/2}$:

$$t_{1/2} = \frac{[A]_0}{2k} = \frac{\frac{1}{2}\left[\left(\frac{0.25\ M}{90\ s}\right)(110\ s) + 1.75\ M\right]}{2\left(\frac{0.25\ M}{90\ s}\right)} = 185\ s$$

S18-6. We analyze the data given below:

TRIAL	[A]	[B]	[C]	INITIAL RATE ($M\ s^{-1}$)
1	0.10	0.10	0.10	0.484
2	0.20	0.10	0.10	0.483
3	0.10	0.20	0.10	0.967
4	0.10	0.10	0.20	1.936

(a) Take each trial in turn.

1. The rate remains effectively the same when [A] is doubled and both [B] and [C] are held constant (trial 2 versus trial 1):

$$\frac{0.483\ M\ s^{-1}}{0.484\ M\ s^{-1}} = 1.00$$

2. The rate doubles when [B] is doubled and both [A] and [C] are held constant (trial 3 versus trial 1):

$$\frac{0.967\ M\ s^{-1}}{0.484\ M\ s^{-1}} = 2.00$$

3. The rate quadruples when [C] is doubled and both [A] and [B] are held constant (trial 4 versus trial 1):

$$\frac{1.936\ M\ s^{-1}}{0.484\ M\ s^{-1}} = 4.00$$

From observation 1 we find a zeroth-order dependence on [A],

$$\text{Rate} \propto [\text{A}]^0$$

whereas from observation 2 we find a first-order dependence on [B],

$$\text{Rate} \propto [\text{B}]^1$$

and from observation 3 we find a second-order dependence on [C]:

$$\text{Rate} \propto [\text{C}]^2$$

(b) Overall, the reaction is third order:

$$\text{Rate} = k[\text{A}]^0[\text{B}]^1[\text{C}]^2 = k[\text{B}][\text{C}]^2$$

$$\text{Order} = 0 + 1 + 2 = 3$$

S18-7. Here we have a reaction

$$2\text{A} \rightarrow \text{products}$$

that obeys a second-order rate law:

$$\text{Rate} = k[\text{A}]^2$$

The reciprocal concentration, 1/[A], varies linearly with time:

$$\frac{1}{[\text{A}]_t} = kt + \frac{1}{[\text{A}]_0}$$

(a) From the initial half-life,

$$t_{1/2} = \frac{1}{k[\text{A}]_0}$$

we first calculate the rate constant:

$$k = \frac{1}{t_{1/2}[\text{A}]_0} = \frac{1}{(155.1\ \text{s})(2.00\ M)} = 0.00322\ M^{-1}\ \text{s}^{-1}$$

Then, rearranging the second-order profile to solve directly for $[A]_t$, we determine the concentration at the specified time:

$$[A]_t = \frac{[A]_0}{kt[A]_0 + 1} = \frac{2.00\ M}{(0.00322\ M^{-1}\ s^{-1})(30.0\ s)(2.00\ M) + 1} = 1.68\ M$$

(b) A bimolecular collision would indeed produce the second-order rate law

$$\text{Rate} = k[A]^2$$

so it is clearly possible for the reaction to be elementary. But other mechanisms, too, may yield the same kinetics, such as the two-step process shown below (one of many conceivable routes):

$$2A \rightarrow I \qquad \text{(slow)}$$

$$I \rightarrow \text{products} \qquad \text{(fast)}$$

Without any experimental studies of the mechanism, we have no way to describe how the reaction actually takes place.

S18-8. The problem stipulates only that a hypothetical reaction

$$A + B \rightarrow C$$

is initially second order in [A]:

$$\text{Rate}(t \approx 0) = k[A]^2$$

(a) No, the initial rate law need not be maintained at all times. Both the form of the rate equation and the value of the rate constant may change as products begin to build up. The concentration of C, for example, may eventually influence the kinetics as the reaction proceeds.

(b) No, this reaction cannot be elementary. A bimolecular collision would produce the rate law

$$\text{Rate} = k[A][B]$$

rather than the observed form $k[A]^2$.

S18-9. We know the activation energies of two reactions, and nothing else:

$$E_{a1} = 100\ \text{kJ mol}^{-1} \qquad E_{a2} = 10\ \text{kJ mol}^{-1}$$

(a) By itself, the activation energy is insufficient to fix the value of either a single rate constant or the ratio of two rate constants. If we assume that the Arrhenius equation is valid for each reaction,

$$k_1 = A_1 \exp\left(-\frac{E_{a1}}{RT}\right)$$

$$k_2 = A_2 \exp\left(-\frac{E_{a2}}{RT}\right)$$

then we also need the ratio of pre-exponential factors (A_1/A_2) to calculate the ratio k_1/k_2:

$$\frac{k_1}{k_2} = \frac{A_1}{A_2} \exp\left(\frac{E_{a2} - E_{a1}}{RT}\right)$$

(b) No. The activation energy, a kinetic quantity, tells us the difference in energy between the reactants and a transition state. To determine whether a reaction is exothermic or endothermic, we need to know the difference in energy (enthalpy) between reactants and final products.

(c) No, the information provided is insufficient to establish the equilibrium constant. We need to know either the standard difference in free energy between reactants and products,

$$K = \exp\left(-\frac{\Delta G^\circ}{RT}\right)$$

or the applicable ratios of the forward and reverse rate constants:

$$K = \frac{k_1}{k_{-1}} \times \frac{k_2}{k_{-2}} \times \cdots \times \frac{k_n}{k_{-n}}$$

S18-10. The Arrhenius equation models the rate constant as a function of temperature:

$$k(T) = A \exp\left(-\frac{E_a}{RT}\right)$$

Both the pre-exponential factor A and the activation energy E_a are positive numbers.

(a) The rate constant increases with temperature:

$$\frac{k(T_2)}{k(T_1)} = \exp\left[\frac{E_a}{R}\left(\frac{1}{T_1} - \frac{1}{T_2}\right)\right] > 1 \qquad \text{if} \qquad T_2 > T_1$$

(b) Take the logarithm of the equation just derived,

$$\ln\frac{k(T_2)}{k(T_1)} = \frac{E_a}{R}\left(\frac{1}{T_1} - \frac{1}{T_2}\right) \equiv \ln p$$

and solve for $1/T_2$:

$$\frac{1}{T_2} = \frac{1}{T_1} - \frac{R\ln p}{E_a}$$

S18-11. Carry over the equation derived in Exercise S18-10 for the ratio p:

$$\frac{1}{T_2} = \frac{1}{T_1} - \frac{R\ln p}{E_a} \qquad\qquad p = \frac{k(T_2)}{k(T_1)}$$

(a) Substitute the values specified for p, E_a, and T_1:

$$p = 3 \qquad E_a = 10{,}000 \text{ J mol}^{-1} \qquad T_1 = 300 \text{ K}$$

$$\frac{1}{T_2} = \frac{1}{300 \text{ K}} - \frac{(8.3145 \text{ J mol}^{-1} \text{ K}^{-1})\ln 3}{10{,}000 \text{ J mol}^{-1}} = 0.0024199 \text{ K}^{-1}$$

$$T_2 = \frac{\text{K}}{0.0024199} = 413.24 \text{ K} = 413 \text{ K} \quad \text{(3 sig fig)}$$

(b) Similar, but with a tenfold greater activation energy:

$$p = 3 \qquad E_a = 100{,}000 \text{ J mol}^{-1} \qquad T_1 = 300 \text{ K}$$

$$\frac{1}{T_2} = \frac{1}{300 \text{ K}} - \frac{(8.3145 \text{ J mol}^{-1} \text{ K}^{-1})\ln 3}{100{,}000 \text{ J mol}^{-1}} = 0.0032420 \text{ K}^{-1}$$

$$T_2 = \frac{\text{K}}{0.0032420} = 308.45 \text{ K} = 308 \text{ K} \quad \text{(3 sig fig)}$$

S18-12. The half-life of a first-order reaction is inversely proportional to the rate constant:

$$t_{1/2} = \frac{\ln 2}{k}$$

(a) Insert $k = 0.500\ \text{s}^{-1}$ to evaluate the half-life at 400 K:

$$t_{1/2} = \frac{0.6931}{0.500\ \text{s}^{-1}} = 1.39\ \text{s}$$

(b) Concentration decreases exponentially in a first-order reaction:

$$[\text{A}]_t = [\text{A}]_0 \exp(-kt) = (0.750\ M)\exp\left[-(0.500\ \text{s}^{-1})(3.25\ \text{s})\right] = 0.148\ M$$

(c) The rate constant presumably increases with temperature, as we have already shown in Exercise S18-10:

$$\frac{k(T_2)}{k(T_1)} = \exp\left[\frac{E_\text{a}}{R}\left(\frac{1}{T_1} - \frac{1}{T_2}\right)\right] > 1 \qquad \text{if} \qquad T_2 > T_1$$

If so, then the half-life will decrease in inverse proportion:

$$t_{1/2} = \frac{\ln 2}{k}$$

(d) The rate constant and half-life of a first-order reaction are independent of the initial concentration.

S18-13. We assume, as in the preceding exercise, that the Arrhenius law applies to a particular first-order reaction:

$$A = 1.00 \times 10^{12}\ \text{s}^{-1} \qquad\qquad E_\text{a} = 50.0\ \text{kJ mol}^{-1} = 5.00 \times 10^4\ \text{J mol}^{-1}$$

(a) Substitute the Arrhenius equation

$$k(T) = A\exp\left(-\frac{E_\text{a}}{RT}\right)$$

into the formula for first-order half-life:

$$t_{1/2} = \frac{\ln 2}{k} = \frac{\ln 2}{A\exp\left(-\frac{E_\text{a}}{RT}\right)} = \frac{0.6931}{(1.00 \times 10^{12}\ \text{s}^{-1})\exp\left[-\frac{5.00 \times 10^4\ \text{J mol}^{-1}}{(8.3145\ \text{J mol}^{-1}\ \text{K}^{-1})(500\ \text{K})}\right]}$$

$$= 1.16 \times 10^{-7}\ \text{s}$$

(b) Similar, but for a higher temperature. The rate constant increases, and the half-life decreases:

$$t_{1/2} = \frac{\ln 2}{k} = \frac{\ln 2}{A \exp\left(-\frac{E_a}{RT}\right)} = \frac{0.6931}{(1.00 \times 10^{12}\ \text{s}^{-1}) \exp\left[-\frac{5.00 \times 10^4\ \text{J mol}^{-1}}{(8.3145\ \text{J mol}^{-1}\ \text{K}^{-1})(600\ \text{K})}\right]}$$

$$= 1.56 \times 10^{-8}\ \text{s}$$

S18-14. The same reaction may proceed at different rates when carried out in different states of matter. We have no grounds to expect that **(a)** the rate law, **(b)** the rate constant, or **(c)** the mechanism will be the same in both solution and gas phases.

S18-15. At equilibrium, the principle of detailed balance provides that

$$k_+[\text{A}] = k_-[\text{B}]$$

for the elementary reaction

$$\text{A} \rightleftarrows \text{B}$$

The forward and reverse rates are equal, with the ratio of k_+ to k_- given by the law of mass action:

$$\frac{k_+}{k_-} = \frac{[\text{B}]}{[\text{A}]} = K$$

Now when the equilibrium constant K is large, as here, the rate constant k_- for the reverse reaction is necessarily small:

$$k_- = \frac{k_+}{K} = 10^{-50}\, k_+$$

Hence the reverse reaction is governed by an exceedingly small rate constant, but the concentration of product is sufficiently large to ensure that the forward and reverse rates are equal.

S18-16. We are given kinetic data for two unimolecular reactions:

REACTION	k_+ (s^{-1})	k_- (s^{-1})
$\text{A} \rightleftarrows \text{B}$	10	20
$\text{C} \rightleftarrows \text{D}$	100	50

(a) According to the principle of detailed balance, the equilibrium constant K is expressed as the ratio k_+/k_-:

$$K_{AB} = \frac{k_{AB+}}{k_{AB-}} = \frac{10\ s^{-1}}{20\ s^{-1}} = 0.50$$

$$K_{CD} = \frac{k_{CD+}}{k_{CD-}} = \frac{100\ s^{-1}}{50\ s^{-1}} = 2.0$$

(b) The overall rate of any reaction at equilibrium is zero.

S18-17. Once more, we apply the principle of detailed balance.

(a) At equilibrium, the overall rate of each reaction—of *any* reaction—is zero.

(b) Given just the equilibrium constant for an elementary reaction, we can determine only the ratio of rate constants,

$$K = \frac{k_+}{k_-}$$

not the numerical values of k_+ and k_- individually.

(c) For an n-step reaction, the principle of detailed balance gives us only the following thermodynamic–kinetic relationship:

$$K = \frac{k_1}{k_{-1}} \times \frac{k_2}{k_{-2}} \times \cdots \times \frac{k_n}{k_{-n}}$$

The value of K, by itself, is insufficient to establish the value of an individual rate constant.

S18-18. We assume that two hypothetical processes share a common set of reactants:

REACTION	ΔG° (kJ mol^{-1})	$\Delta G^\ddagger$ (kJ mol^{-1})
$R \rightleftarrows P_1$	−100	100
$R \rightleftarrows P_2$	−10	10

(a) The rate of reaction is affected by the difference in free energy between the reactants and a transition state, or activated complex. Formation of an activated complex is favored in the process with the lower (less positive) free energy of activation:

$$\text{R} \rightleftarrows \text{P}_2 \qquad \Delta G^{\ddagger} = 10 \text{ kJ mol}^{-1}$$

Reaction 2 will therefore proceed faster than reaction 1.

(b) The equilibrium constant depends on the difference in free energy between the reactants and the final products:

$$K = \exp\left(-\frac{\Delta G^{\circ}}{RT}\right)$$

Process 1, which has a more negative free energy of reaction than process 2, yields the higher proportion of products at equilibrium:

$$\text{R} \rightleftarrows \text{P}_1 \qquad \Delta G^{\circ} = -100 \text{ kJ mol}^{-1}$$

S18-19. A continuation of Exercise S18-18.

(a) The product P_1 ($\Delta G^{\circ} = -100$ kJ mol^{-1}) is thermodynamically more stable than the product P_2 ($\Delta G^{\circ} = -10$ kJ mol^{-1}). If the three-way reaction comes to a true equilibrium,

$$\text{P}_1 \rightleftarrows \text{R} \rightleftarrows \text{P}_2$$

then P_1 will be present in the greater proportion.

(b) Initially, the process is controlled by kinetics: The product that can be produced faster (P_2) appears first. In the long run, however, the thermodynamically more stable product (P_1) will dominate—*if* equilibrium can be established.

S18-20. One last look at the transformations introduced in the preceding two exercises.

(a) If P_2 is removed from the system, then the three-way process

$$\text{P}_1 \rightleftarrows \text{R} \rightleftarrows \text{P}_2$$

cannot come to a true equilibrium. All three species—P_1, P_2, and R—must be present for equilibrium to be attained.

(b) The process, as described, is controlled by kinetics: The kinetically favored product (P_2) is produced first and subsequently disappears from the system, permanently depleting the stock of reactants. The thermodynamically stable product (P_1) is denied the opportunity to form.

S18-21. The overall reaction

$$A + 2B + C \rightarrow D$$

probably proceeds through an intermediate (or series of intermediates) during its initial stages. The appearance of D would then be delayed by the production and consumption of the intermediate species.

S18-22. The transition state, existing in a condition of unstable equilibrium, should appear at a local maximum of potential energy—at a point where the tangent to the curve is horizontal, like this:

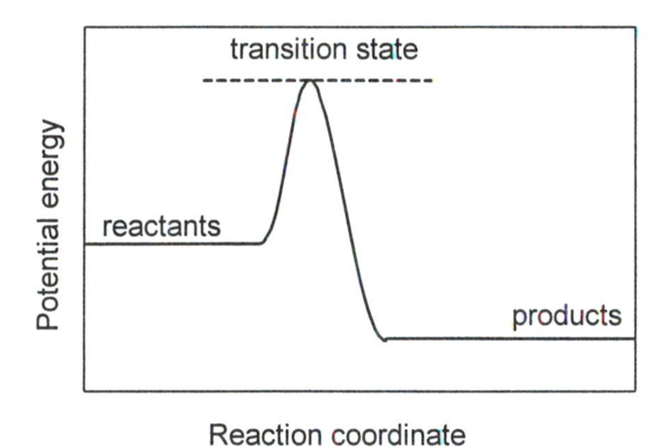

Not like this:

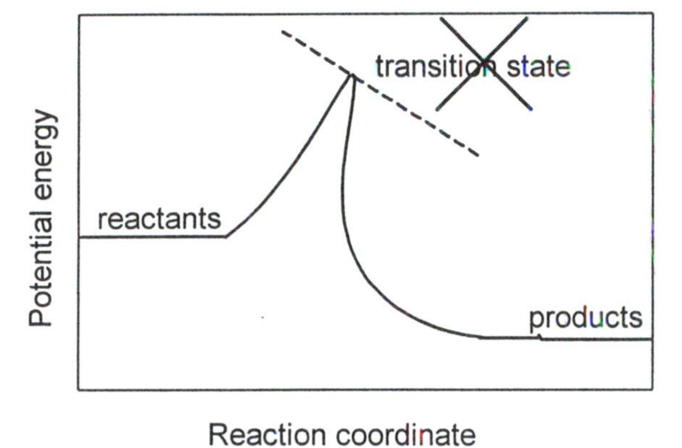

Chapter 19

Chemistry Coordinated—The Transition Metals and Their Complexes

S19-1. Analyze the coordination compound $K_2[Ni(en)Cl_4]$ as follows: **(a)** Identify the coordinated metal and its oxidation state. **(b)** Identify the ligands and deduce the coordination number. **(c)** Describe the geometric structure.

S19-2. Analyze the coordination compound $[Co(en)_3]Cl_3$ as follows: **(a)** Identify the coordinated metal and its oxidation state. **(b)** Identify the ligands and deduce the coordination number. **(c)** Describe the geometric structure.

S19-3. **(a)** Which one of these two complex ions—$[Co(NH_3)_6]^{3+}$ or $[CoF_6]^{3-}$—is likely to be paramagnetic? Which one is likely to be diamagnetic? **(b)** For each structure, use crystal field theory to predict the electron configuration.

S19-4. The complex ion $[AuCl_4]^-$ is square planar. **(a)** What is the oxidation state of the gold ion? **(b)** How many *d* electrons are present? **(c)** Is the complex ion paramagnetic or diamagnetic?

S19-5. **(a)** Suppose that a material absorbs light broadly at wavelengths between 360 nm and 780 nm. What color—red, green, violet, gray-black, or no color at all—do you expect the material to have? **(b)** What color do you expect for a material that absorbs light primarily at a wavelength of 300 nm? **(c)** What color do you expect for a material that absorbs light primarily at a wavelength of 850 nm?

S19-6. An aqueous solution of $[Co(H_2O)_6]^{2+}$ absorbs light at a frequency of 5.77×10^{14} Hz. **(a)** Calculate the wavelength of absorption. In which portion of the electromagnetic spectrum—infrared, visible, or ultraviolet—does the radiation fall? **(b)** Calculate the crystal field splitting parameter Δ, expressing your answer in kJ mol^{-1}.

S19-7. Which one of these two complexes—$[Co(NH_3)_6]^{3+}$ or $[Co(NH_3)_5H_2O]^{3+}$—absorbs light at the longer wavelength?

S19-8. Which one of these two complexes—$[Cd(NH_3)_6]^{2+}$ or $[Co(NH_3)_6]^{2+}$—is colorless?

S19-9. Is $[Zn(H_2O)_4]^{2+}$ colored or colorless?

S19-10. Solutions containing the complex ion $[Co(NH_3)_6]^{3+}$ are yellow. In which range of wavelengths (see below) do you expect the structure to absorb light?

400 nm – 450 nm　　　500 nm – 550 nm　　　600 nm – 650 nm

S19-11. How many *d* electrons are present in the Sc^{3+} ion? Do you expect an octahedral complex built around Sc^{3+} to be especially stable or unstable?

S19-12. **(a)** How many *d* electrons are present in Ti^{4+} and Mn^{7+}? **(b)** Which one of the two ions do you expect to be more stable? **(c)** Which one of these two species—Mn^{7+} or MnO_4^-—do you expect to be more stable?

S19-13. The ions Zn^{2+} and Ag^+ are especially stable. Explain why.

S19-14. Which one of these two complexes—$[Ag(NH_3)_2]^+$ or $[Ag(CN)_2]^-$—is thermodynamically more stable? Relevant data may be found in Appendix C of *Principles of Chemistry.*

S19-15. Calculate $\Delta G°$ for the formation of $[PbCl_4]^{2-}$:

$$Pb^{2+}(aq) + 4Cl^-(aq) \rightleftarrows [PbCl_4]^{2-}(aq)$$

Relevant data may be found in Appendix C.

S19-16. Calculate the equilibrium concentrations of Ag^+ and $S_2O_3^{2-}$ that result when 0.500 mol $[Ag(S_2O_3)_2]^{3-}$ is dissolved in 0.250 L H_2O. See the reaction below:

$$Ag^+(aq) + 2S_2O_3^{2-}(aq) \rightleftarrows [Ag(S_2O_3)_2]^{3-}(aq) \qquad K_f = 2.9 \times 10^{13}$$

S19-17. Use the following questions to draw an analogy between the heterogeneous equilibrium of an ionic solute AB_n,

$$AB_n(s) \rightleftarrows A^{n+}(aq) + nB^-(aq)$$

and the equilibrium between a coordination complex ML_n and its ligands:

$$ML_n(aq) \rightleftarrows M(aq) + nL(aq)$$

(a) Write an expression for the equilibrium constant in each system. **(b)** Which is the ordered subsystem and which is the disordered subsystem in each transformation? **(c)** Does your general picture apply equally well to the dissolution–precipitation equilibrium of a molecular solute X? See below:

$$X(s) \rightleftarrows X(aq)$$

S19-18. The formation constant for $[Zn(OH)_4]^{2-}$ is 2.8×10^{15} in aqueous solution. The formation constant for $[Zn(NH_3)_4]^{2+}$ is 2.9×10^9. **(a)** Can you predict which of the two complex ions will form faster? **(b)** Suppose that ammonia molecules and hydroxide ions compete simultaneously for the same zinc ions. Can you predict which one of the two complex ions will predominate at equilibrium?

S19-19. **(a)** Explain the difference between a thermodynamically *stable* complex and a kinetically *inert* complex. **(b)** Which quantity—the free energy of formation or the free energy of activation—determines stability and instability? **(c)** Which quantity determines inertness and lability?

S19-20. Octahedral complexes built around Cr^{3+} generally are both thermodynamically stable and kinetically inert compared with those built around Cr^{2+}. Explain why.

S19-21. Nearly all complexes of Cr^{3+} are octahedral. Explain why tetrahedral and square planar complexes are unfavorable.

S19-22. Do you expect Cr^{2+} to be a reducing agent or an oxidizing agent?

S19-23. Suppose that a certain complex is soluble in both water and ethanol. **(a)** Will the formation constant necessarily be the same? **(b)** Will the lability of the complex necessarily be the same? Explain why or why not.

SOLUTIONS

S19-1. Here the oxidation state of the metal plus the sum of all ionic charges—from both the ligands and the counterions—must equal zero, the net charge of a neutral coordination compound.

(a) One ethylenediamine molecule and four Cl^- ions are coordinated around Ni^{2+} in the complex ion $[Ni(en)Cl_4]^{2-}$, a doubly negative species. Two K^+ counterions outside the coordination sphere ensure that the coordination compound $K_2[Ni(en)Cl_4]$ is neutral:

$$Ni^{2+} + en + 4Cl^- \rightarrow [Ni(en)Cl_4]^{2-}$$

$$2K^+ + [Ni(en)Cl_4]^{2-} \rightarrow K_2[Ni(en)Cl_4]$$

(b) There are four monodentate ligands (Cl^-) and one bidentate ligand (en), corresponding to a coordination number of 6. The bidendate ligand occupies two positions in the coordination sphere.

(c) The complex ion is octahedral. For every $[Ni(en)Cl_4]^{2-}$ anion, there are two K^+ counterions outside.

S19-2. $[Co(en)_3]Cl_3$ consists of the complex cation $[Co(en)_3]^{3+}$ and three Cl^- counterions.

(a) Three ethylenediamine molecules are coordinated around Co^{3+} in the complex ion $[Co(en)_3]^{3+}$. Three Cl^- counterions lie outside the coordination sphere, producing the neutral compound $[Co(en)_3]Cl_3$:

$$Co^{3+} + 3en \rightarrow [Co(en)_3]^{3+}$$

$$[Co(en)_3]^{3+} + 3Cl^- \rightarrow [Co(en)_3]Cl_3$$

(b) There are three bidentate ligands (en), corresponding to a coordination number of 6.

(c) The complex ion is octahedral. For every $[Co(en)_3]^{3+}$ cation, there are three Cl^- counterions outside.

S19-3. Cobalt exists as Co^{3+} (a d^6 ion) in each of the complexes $[Co(NH_3)_6]^{3+}$ and $[CoF_6]^{3-}$. An octahedral d^6 species will be paramagnetic if its configuration is high spin,

$$\underline{\uparrow}\ \underline{\uparrow}\quad e_g$$

$$\underline{\uparrow\downarrow}\ \underline{\uparrow}\ \underline{\uparrow}\quad t_{2g}$$

and diamagnetic if its configuration is low spin:

$$\underline{\ \ }\ \underline{\ \ }\quad e_g$$

$$\underline{\uparrow\downarrow}\ \underline{\uparrow\downarrow}\ \underline{\uparrow\downarrow}\quad t_{2g}$$

(a) The weak-field ligand, F^-, is more likely to produce a high-spin configuration: $[CoF_6]^{3-}$ is expected to be paramagnetic. The strong-field ligand, NH_3, is expected to produce a low-spin, diamagnetic complex: $[Co(NH_3)_6]^{3+}$.

(b) Read the configurations off the spin diagrams previously shown:

$[CoF_6]^{3-}$ $(t_{2g})^4(e_g)^2$

$[Co(NH_3)_6]^{3+}$ $(t_{2g})^6$

S19-4. Four Cl^- ligands are coordinated around Au^{3+} in the complex ion $[AuCl_4]^-$:

$$Au^{3+} + 4Cl^- \rightarrow [AuCl_4]^-$$

(a) See above: Gold exists as Au^{3+} in the anionic complex. The net charge is –1.

(b) The configuration of neutral Au is $[Xe]6s^14f^{14}5d^{10}$, and the configuration of Au^{3+} is $[Xe]4f^{14}5d^8$. One 6*s* and two 5*d* electrons are lost, leaving the ion with eight *d* electrons.

(c) Crystal field theory predicts that a square planar d^8 complex will be diamagnetic. All the electrons are paired:

$x^2 - y^2$ —

xy ↑↓

z^2 ↑↓

xz, yz ↑↓ ↑↓

Note that the splittings above are not drawn to scale.

S19-5. If a structure absorbs visible light of a particular color, then it *transmits* a spectrum of wavelengths corresponding to the complementary color.

(a) A material that absorbs broadly over the entire visible spectrum (fully covered by the range 360 nm – 780 nm) is expected to be gray or black. No visible light comes through.

(b) A material that absorbs light at 300 nm (ultraviolet) is expected to be colorless. No visible wavelengths are lost in transmission.

(c) A material that absorbs light in the infrared (850 nm) is expected to be colorless as well. No visible wavelengths are lost in transmission.

S19-6. We simply apply the equations $\lambda\nu = c$ and $E = h\nu$.

(a) Wavelength is inversely proportional to frequency:

$$\lambda = \frac{c}{\nu} = \frac{2.998 \times 10^{8}\ \text{m s}^{-1}}{5.77 \times 10^{14}\ \text{s}^{-1}} = 5.20 \times 10^{-7}\ \text{m} = 520\ \text{nm} \quad \text{(visible)}$$

(b) Convert joules per photon into kilojoules per mole:

$$\Delta = E = h\nu = \left(6.626 \times 10^{-34}\ \text{J s}\right)\left(5.77 \times 10^{14}\ \text{s}^{-1}\right) \times \frac{6.022 \times 10^{23}}{\text{mol}} \times \frac{1\ \text{kJ}}{1000\ \text{J}} = 230.\ \text{kJ mol}^{-1}$$

S19-7. Since H_2O lies below NH_3 in the spectrochemical series,

$$I^- < Br^- < Cl^- < F^- < OH^- < H_2O < NCS^- < NH_3 < \text{en} < CO, CN^-$$

weak field intermediate field strong field

the complex $[Co(NH_3)_5H_2O]^{3+}$ is expected to absorb light at a longer wavelength (lower energy) than the complex $[Co(NH_3)_6]^{3+}$.

S19-8. The energy of visible light typically falls within the range needed to promote an electron from an occupied to an unoccupied *d* orbital. Thus the complex $[Cd(NH_3)_6]^{2+}$, built around Cd^{2+} (a d^{10} species), is colorless: No transition is possible within the manifold of filled *d* levels. By contrast, cobalt exists as Co^{2+} (a d^7 species) in $[Co(NH_3)_6]^{2+}$.

S19-9. Similar: Zinc exists as Zn^{2+}, a d^{10} ion, in $[Zn(H_2O)_4]^{2+}$. The complex is colorless.

S19-10. A material that appears yellow will absorb light primarily in the violet region of the spectrum (the color complementary to yellow). The range of wavelengths is most likely between 400 nm and 450 nm.

S19-11. A neutral scandium atom has the configuration $[Ar]4s^23d^1$, and the tripositive Sc^{3+} ion has the noble-gas configuration [Ar]. Complexes built around Sc^{3+} are especially stable as a result. Repulsive interactions, normally present between lone pairs on the ligands and *d* electrons on the metal, are avoided.

S19-12. A comparison of Ti^{4+} and Mn^{7+}—as free ions and also in complexes.

(a) Both Ti^{4+} and Mn^{7+} are d^0 ions isoelectronic with argon.

(b) Although each ion has a noble-gas configuration, the stability of free Mn^{7+} is problematic owing to the high charge. The Ti^{4+} ion is more stable by comparison.

(c) An MnO_4^- ion, in which Mn^{7+} is coordinated to four (negative) oxide ions, is more stable than Mn^{7+} alone. The O^{2-} ligands stabilize the system both by reducing the net charge and by forming bonds with manganese.

S19-13. Neutral zinc and silver atoms are configured as $4s^2 3d^{10}$ and $5s^1 4d^{10}$, respectively. Zn^{2+} and Ag^+ ions are both d^{10} species and thus gain stability from their closed subshells.

S19-14. Table C-22 in the text shows that the formation constant is substantially larger for $[Ag(CN)_2]^-$ than for $[Ag(NH_3)_2]^+$:

$$Ag^+(aq) + 2CN^-(aq) \rightleftarrows [Ag(CN)_2]^-(aq) \qquad K_f = 1.0 \times 10^{21}$$

$$Ag^+(aq) + 2NH_3(aq) \rightleftarrows [Ag(NH_3)_2]^+(aq) \qquad K_f = 1.7 \times 10^{7}$$

The cyanide complex is thermodynamically more stable.

S19-15. From the formation constant listed in Table C-22,

$$Pb^{2+}(aq) + 4Cl^-(aq) \rightleftarrows [PbCl_4]^{2-}(aq) \qquad K_f = 25$$

we calculate the standard difference in free energy:

$$\Delta G^\circ = -RT \ln K_f = -\left(8.3145 \times 10^{-3}\ \text{kJ mol}^{-1}\ \text{K}^{-1}\right)\left(298.15\ \text{K}\right)\ln 25 = -8.0\ \text{kJ mol}^{-1}$$

S19-16. Start with the formation equilibrium,

$$Ag^+(aq) + 2S_2O_3^{2-}(aq) \rightleftarrows [Ag(S_2O_3)_2]^{3-}(aq) \qquad K_f = 2.9 \times 10^{13}$$

and then turn the equation around:

$$[Ag(S_2O_3)_2]^{3-}(aq) \rightleftarrows Ag^+(aq) + 2S_2O_3^{2-}(aq) \qquad K = \frac{1}{K_f} = 3.45 \times 10^{-14}$$

Initial conc.	$\dfrac{0.500\ \text{mol}}{0.250\ \text{L}} = 2.00\ M$	0	0
Change	$-x$	x	$2x$
Equil. conc.	$2.00 - x$	x	$2x$

Solution of the mass-action equation

$$K = \frac{[Ag^+][S_2O_3^{2-}]^2}{[Ag(S_2O_3)_2^{3-}]} = \frac{x(2x)^2}{2.00 - x} = \frac{4x^3}{2.00 - x} = 3.45 \times 10^{-14}$$

is simplified by the assumption that $2.00 - x \approx 2.00$:

$$\frac{4x^3}{2} \approx 3.45 \times 10^{-14}$$

$$x = [Ag^+] = 2.6 \times 10^{-5}\ M \quad \text{(2 sig fig)}$$

$$2x = [S_2O_3^{2-}] = 5.2 \times 10^{-5}\ M$$

S19-17. Order and disorder.

(a) The equilibria are analogous. The expressions for K differ only superficially—in the conventional absence of solid solute from the solubility-product constant:

$$AB_n(s) \rightleftarrows A^{n+}(aq) + nB^-(aq) \qquad K = [A^{n+}][B^-]^n$$

$$ML_n(aq) \rightleftarrows M(aq) + nL(aq) \qquad K = \frac{[M][L]^n}{[ML_n]}$$

(b) For the dissolution of AB_n, the aqueous phase is disordered relative to the solid phase. For the dissociation of ML_n, the dissociated complex is disordered relative to the associated complex.

(c) The same general principles hold for the heterogeneous equilibrium of a molecular solute in a saturated solution.

S19-18. We compare the formation of $[Zn(OH)_4]^{2-}$ with the formation of $[Zn(NH_3)_4]^{2+}$:

$$Zn^{2+}(aq) + 4OH^-(aq) \rightleftarrows [Zn(OH)_4]^{2-}(aq) \qquad K_f = 2.8 \times 10^{15}$$

$$Zn^{2+}(aq) + 4NH_3(aq) \rightleftarrows [Zn(NH_3)_4]^{2+}(aq) \qquad K_f = 2.9 \times 10^9$$

(a) The formation constants are thermodynamic parameters. Reflecting a difference in free energy between reactants and final products, they tell us nothing about the activation energy (and hence nothing about the speed of reaction).

(b) The complex with the larger formation constant—$[Zn(OH)_4]^{2-}$—is thermodynamically more stable and is expected to predominate at equilibrium.

S19-19. Thermodynamics versus kinetics, again.

(a) A thermodynamically stable complex is characterized by a large formation constant and a correspondingly high proportion of fully associated structures at equilibrium. A kinetically inert complex is simply a complex that is slow to react and exchange ligands, regardless of the size of its formation constant.

(b) Thermodynamic stability is determined by the free energy of reaction—the difference ΔG° between reactants and products.

(c) Kinetic inertness is determined by the free energy of activation—the difference $\Delta G^\ddagger$ between reactants and a transition state.

S19-20. Cr^{3+}, a d^3 ion, enjoys the thermodynamic stability of a half-filled t_{2g} subshell in an octahedral complex:

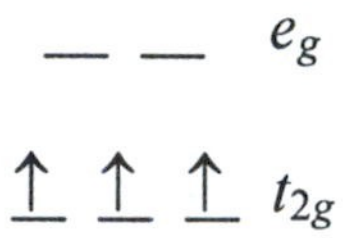

$$\begin{array}{ll} \_\ \_\quad e_g & \\ & CFSE = -1.2\,\Delta \\ \uparrow\ \uparrow\ \uparrow\quad t_{2g} & \end{array}$$

This half-filled $(t_{2g})^3$ configuration is more stable than either the high-spin $(t_{2g})^3(e_g)^1$ configuration or the low-spin $(t_{2g})^4$ configuration available to Cr^{2+}. The high-spin alternative, which places an electron in an unfavorable e_g level, has a more positive *CFSE*:

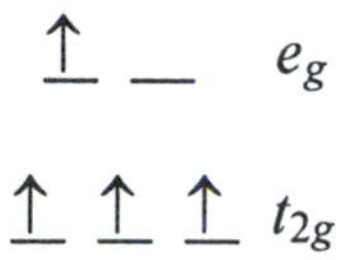

$$\begin{array}{ll} \uparrow\ \_\quad e_g & \\ & CFSE = -0.6\,\Delta \\ \uparrow\ \uparrow\ \uparrow\quad t_{2g} & \end{array}$$

The low-spin alternative, which forces two electrons to pair up in a t_{2g} orbital, brings about increased electron–electron repulsion (despite the enhanced *CFSE*):

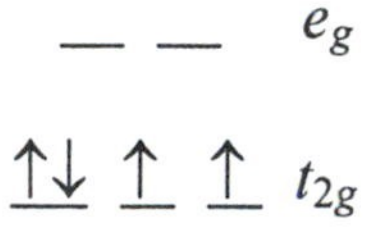

$$\begin{array}{ll} \_\ \_\quad e_g & \\ & CFSE = -1.6\,\Delta \\ \uparrow\downarrow\ \uparrow\ \uparrow\quad t_{2g} & \end{array}$$

Octahedral Cr^{3+} is *slow* to react, too, if the d^3 configuration is altered during the formation of a transition state. The activation barrier is high.

S19-21. Only in an octahedral crystal field does the d^3 configuration of Cr^{3+} produce a half-filled subshell:

octahedral	tetrahedral	square planar
		__ $x^2 - y^2$
		__ xy
__ __ $z^2, x^2 - y^2$	↑ __ __ xy, xz, yz	__ z^2
↑ ↑ ↑ xy, xz, yz	↑ ↑ $z^2, x^2 - y^2$	↑↓ ↑ xz, yz

Splittings are not drawn to scale.

S19-22. Cr^{2+}, a reducing agent, undergoes oxidation to Cr^{3+}:

$$Cr^{2+}(aq) \rightarrow Cr^{3+}(aq) + e^- \qquad \mathcal{E}^\circ_{ox} = 0.41 \text{ V}$$

Doing so, it goes from a d^4 configuration to a stable d^3 configuration. See Exercise S19-20.

S19-23. Thermodynamic and kinetic properties are both affected by the solvent particles, which interact with the complex and also with its separated ligands. Neither **(a)** the formation constant nor **(b)** the degree of lability need be the same in ethanol as they are in water.

Chapter 20

Spectroscopy and Analysis

S20-1. Match the items on the left with the items on the right:

gamma rays	molecular rotation
infrared radiation	nuclear spin
microwave radiation	molecular vibration
ultraviolet/visible radiation	core electrons
X rays	protons and neutrons
radiofrequency radiation	valence electrons

S20-2. Describe (or sketch) the ^{1}H NMR spectrum of each structure proposed below, paying attention to the number of lines and their relative intensities:

(a) HCCH

(b) HCCBr

(c) $CH_3CBr_2CBr_2CH_3$

(d) $CH_3CBr_2CBr_2CCl_2H$

Rewrite each of the formulas to make every connection and every bond explicit.

S20-3. Similar—describe or sketch the ^{1}H NMR spectrum of each proposed structure:

(a) CH_2BrCBr_2H

(b) CH_2Br_2

(c) CH_2CBr_2

(d) *trans*-CHBrCHCl

Rewrite the formulas to make every connection and every bond explicit.

S20-4. Carbon-13, which makes up only 1.1% of naturally occurring carbon, has a nuclear spin quantum number of $\frac{1}{2}$. The abundant ^{12}C isotope, by contrast, has a nuclear spin quantum number of 0. **(a)** Sketch the ^{13}C NMR spectrum of $^{12}CCl_3{}^{13}CH_2{}^{12}CCl_3$, taking into account the local field produced by the hydrogen nuclei. **(b)** How would the ^{13}C spectrum change if the hydrogen nuclei did not contribute a local field?

S20-5. The mass spectrum of chromium contains four signals:

RELATIVE MASS	INTENSITY
49.946046	0.05186
51.940509	1.00000
52.940651	0.11339
53.938882	0.02823

(a) To which isotope does each signal correspond? (Note that the mass of an atomic nucleus is slightly less than the mass of its protons and neutrons taken separately. See Chapter 21.) **(b)** Calculate the average molar mass of chromium, using only the data supplied here. **(c)** Suppose that a naturally occurring sample of chromium contains 20.00 g. How many grams of ^{53}Cr are present in the sample?

S20-6. Sketch the mass spectrum of lead, consulting Table C-11 as needed.

S20-7. The mass spectrum of $C_{18}H_{38}$ contains many signals, including a peak corresponding to a mass/charge ratio of 254. Do you expect this signal to be the most intense in the spectrum? Why or why not?

S20-8. Transitions between rotational energy levels usually occur when an oscillating electric field interacts with a permanent electric dipole moment on a structure. Which of the following species are "microwave-active" (able to produce a rotational spectrum)?

(a) H_2O **(b)** CH_4 **(c)** HCl **(d)** CO_2

(e) NH_3 **(f)** Ar **(g)** CH_3Cl

S20-9. A nonpolar molecule, despite the absence of a permanent electric dipole moment, sometimes can produce a rotational spectrum. To do so, the structure must possess a permanent *magnetic* dipole moment able to interact with a magnetic field. Which one of these diatomic molecules—H_2, Li_2, or O_2—therefore might be microwave-active in its ground state? If necessary, review the discussion of diatomic molecular orbitals in Chapter 7.

S20-10. Transitions between vibrational levels usually occur when a dipole moment *changes* while it interacts with an electric field. Which of the following modes of vibration are "infrared-active" and thus will appear in a vibrational spectrum?

(a) stretching of OH in H_2O
(b) bending of HOH in H_2O
(c) stretching of H_2
(d) stretching of HCl
(e) symmetric stretching of OCO in CO_2
(f) bending of OCO in CO_2

Hint: During the course of a "symmetric stretch," the lengths of the two CO bonds in CO_2 increase and decrease in tandem—equally on either side of the molecule. The OCO bond angle remains unchanged during a symmetric stretching vibration, but the angle varies periodically during a bending vibration.

S20-11. When an excited helium atom undergoes a transition from the $1s^12p^1$ state down to the ground state ($1s^2$), it emits electromagnetic radiation at a wavelength of 58.43 nm—the so-called He(I) line. **(a)** In what portion of the electromagnetic spectrum does the He(I) line fall? **(b)** What is the corresponding frequency? **(c)** What is the corresponding photon energy?

S20-12. Here is an example of *ultraviolet photoelectron spectroscopy*: **(a)** Write the valence electron configuration of N_2, referring back to Chapter 7 as needed. Is the highest occupied molecular orbital (abbreviated HOMO) of σ symmetry or π symmetry? **(b)** The ionization energy of an electron in the HOMO of N_2 is 1504 kJ mol^{-1}. If this electron is ionized by He(I) radiation (see the preceding exercise), with what kinetic energy will it emerge? **(c)** What will be its speed?

S20-13. When excited by a wavelength of 58.43 nm, as above, N_2 also ejects an electron from an orbital lying just below the HOMO. **(a)** The speed of this additional photoelectron is 1.262×10^6 m s^{-1}. What is the corresponding kinetic energy? **(b)** Calculate the ionization energy in kJ mol^{-1}.

S20-14. Do you expect ultraviolet photoelectron spectroscopy, as illustrated in the preceding two examples, to be generally useful for studying electrons in the core of an atom or molecule?

S20-15. The 1*s* ionization energies for three second-row atoms are listed below:

(a) Li 4.82×10^3 kJ mol^{-1}
(b) B 1.83×10^4 kJ mol^{-1}
(c) F 6.66×10^4 kJ mol^{-1}

Calculate the wavelength needed to eject a core electron from each atom. In what portion of the electromagnetic spectrum does each of the energies lie?

SOLUTIONS

S20-1. Portions of the electromagnetic spectrum are presented, from top to bottom, in descending order of photon energy:

gamma rays	protons and neutrons
X rays	core electrons
ultraviolet/visible radiation	valence electrons
infrared radiation	molecular vibration
microwave radiation	molecular rotation
radiofrequency radiation	nuclear spin

S20-2. The stick spectra in this exercise and those following are intended mainly to show the number of lines and their relative intensities. The positions of the lines are schematic, not to be interpreted in any absolute sense.

Hydrogen nuclei in equivalent environments are labeled with the same subscript (for example, H_a).

(a) Both ^{1}H nuclei in acetylene are equivalent:

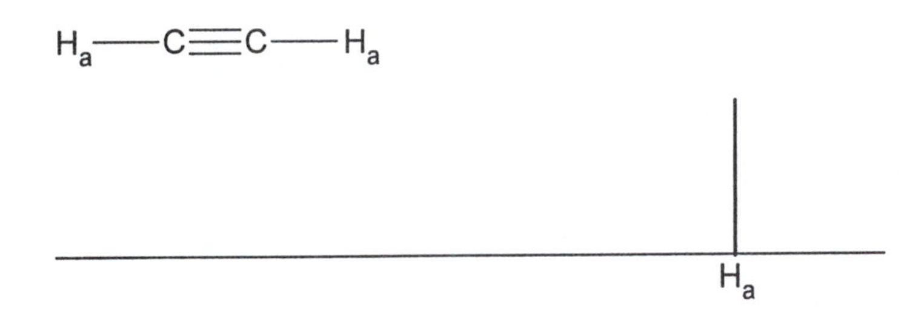

The spectrum contains just one line.

(b) Again, there is just one line—but the environment in bromoacetylene differs from the environment in acetylene. The ^{1}H signal in H–C≡C–Br, influenced by the electron-withdrawing bromine atom, is deshielded relative to its position in H–C≡C–H:

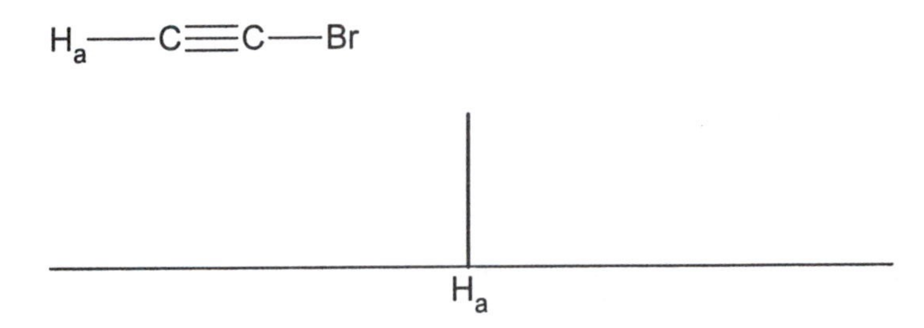

(c) All six protons are equivalent. There is just one line:

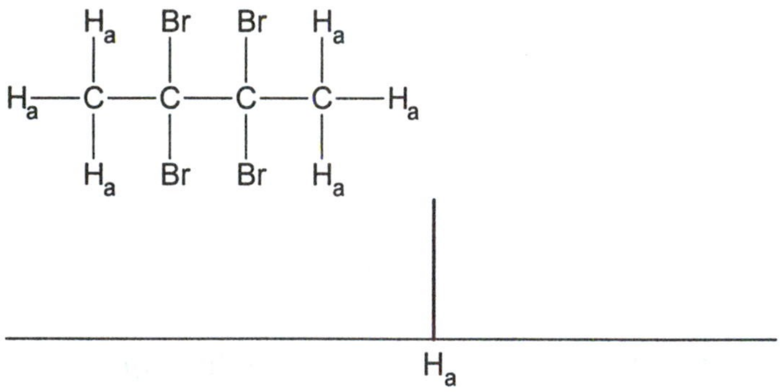

(d) Here we have a group of three equivalent protons (H_a) and a single proton (H_b). The intensity ratio is 3:1, with the CCl_2H proton deshielded relative to the methyl protons:

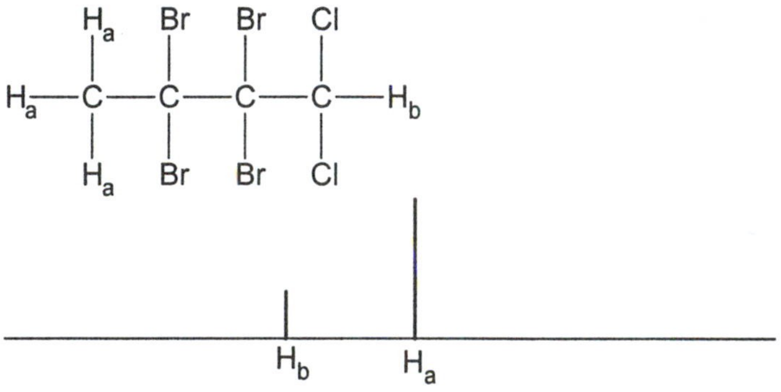

Separated by four carbon atoms, H_a and H_b do not interact appreciably via the J coupling. The signals are not split.

S20-3. More ^{1}H NMR stick spectra.

(a) The spectrum contains five lines, a group of two and a group of three. (1) H_b presents its local field to H_a in two equivalent ways (↑ or ↓), thereby splitting the H_a resonance into a 1:1 doublet:

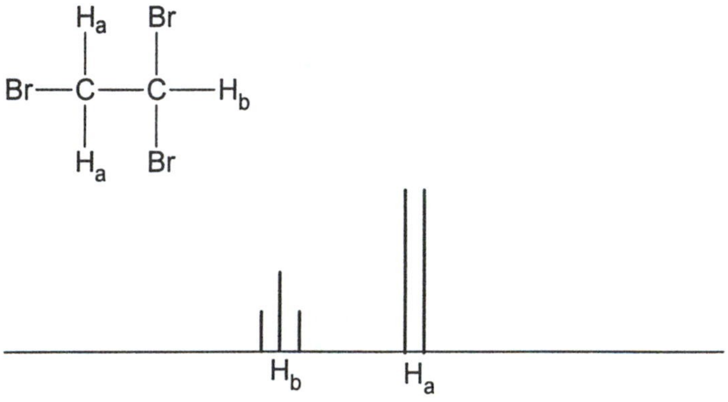

(2) The H_b proton, responding to three quantized fields from the two H_a protons, resonates as a 1:2:1 triplet:

↓↓

↑↓ or ↓↑

↑↑

Coming from just one proton, the combined intensity of the triplet is half the combined intensity of the doublet.

Note, finally, that H_b is deshielded relative to H_a. The proton in the CBr_2H group is strongly influenced by *two* electron-withdrawing bromine atoms, not one.

(b) The two protons are equivalent:

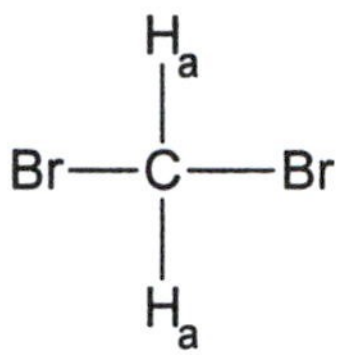

H_a

Br—C—Br

H_a

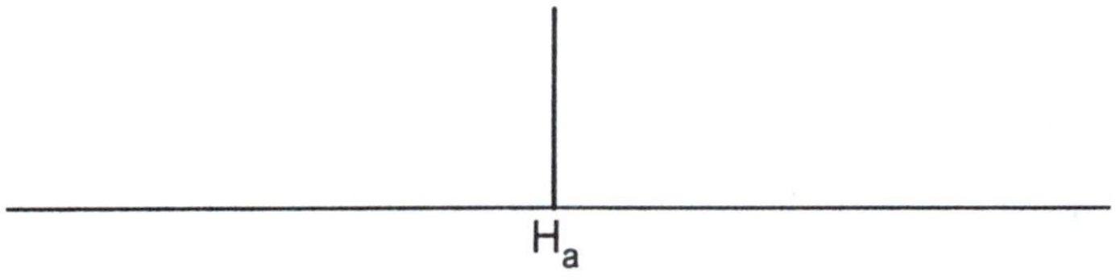

H_a

Only one line is observed, a singlet.

(c) Similar. The two protons in 1,1-dibromoethene are equivalent,

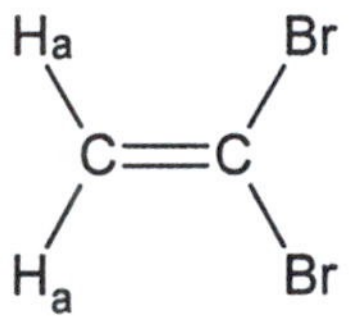

Ha Br

C═C

H_a Br

and consequently the ^1H NMR spectrum contains only one resonance.

(d) Each proton splits the signal of the other into a symmetric doublet. The combined intensity is the same for both:

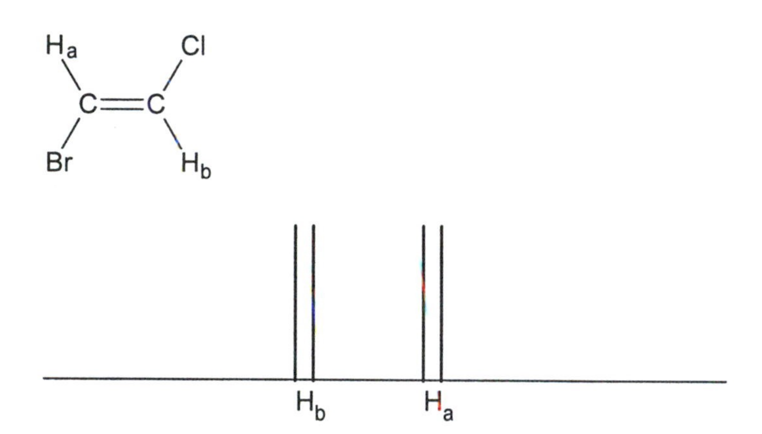

A similar pattern would develop for the cis isomer,

H_a H_b
C=C
Br Cl

although with different chemical shifts for H_a and H_b.

S20-4. Only the central carbon nucleus, specified as ^{13}C, produces an NMR spectrum:

Cl H Cl
Cl—^{12}C—^{13}C$_a$—^{12}C—Cl
Cl H Cl

(a) The carbon-13 signal is split into a 1:2:1 triplet by the local field of the two hydrogen nuclei:

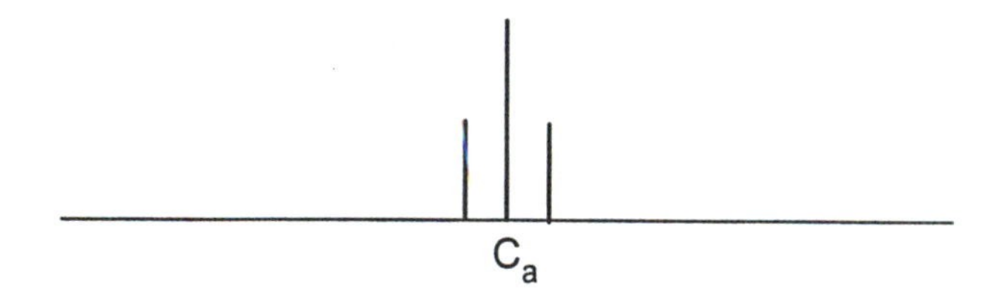

Each component of the triplet corresponds to a particular arrangement of the ^{1}H spins:

$$\downarrow\downarrow$$

$$\uparrow\downarrow \quad \text{or} \quad \downarrow\uparrow$$

$$\uparrow\uparrow$$

(b) The carbon-13 spectrum collapses to a single line once the local field of the hydrogen nuclei is removed:

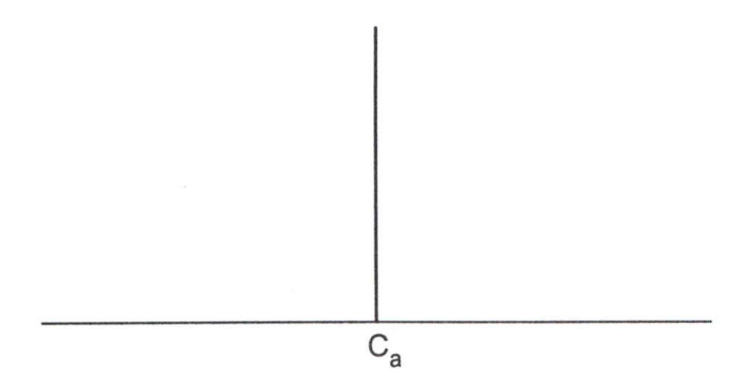

S20-5. Chromium exists as four stable isotopes: ^{50}Cr, ^{52}Cr, ^{53}Cr, and ^{54}Cr.

(a) The mass of each nucleus is slightly less than the mass of its protons and neutrons taken separately:

ISOTOPE	RELATIVE MASS	INTENSITY
^{50}Cr	49.946046	0.05186
^{52}Cr	51.940509	1.00000
^{53}Cr	52.940651	0.11339
^{54}Cr	53.938882	0.02823

(b) The peaks in the mass spectrum are proportional to the relative abundance of the different isotopes. Accordingly, we have the relationship

$$\text{Average mass of chromium} = X_{50}m_{50} + X_{52}m_{52} + X_{53}m_{53} + X_{54}m_{54}$$

where X_A is the mole fraction and m_A is the mass of the isotope ^{A}Cr.

The relative mass of, say, ^{53}Cr, is stated in the problem as 52.940651, and its mole fraction is 0.09501:

$$X_{53} = \frac{0.11339}{0.05186 + 1.00000 + 0.11339 + 0.02823} = 0.09501$$

Calculating all the other mole fractions in the same way, we obtain an average relative mass of 51.996 (average molar mass = 51.996 g mol^{-1}):

$$\mathcal{m} = X_{50}m_{50} + X_{52}m_{52} + X_{53}m_{53} + X_{54}m_{54}$$

$$= \frac{0.05186}{1.19348} \times 49.946046 + \frac{1.00000}{1.19348} \times 51.940509 + \frac{0.11339}{1.19348} \times 52.940651$$

$$+ \frac{0.02823}{1.19348} \times 53.938882 = 51.996$$

(c) Use the mole fraction calculated above in (b):

$$20.00 \text{ g Cr} \times \frac{0.09501 \text{ mol } ^{53}\text{Cr}}{\text{mol Cr}} = 1.900 \text{ g } ^{53}\text{Cr}$$

S20-6. The relative abundances are listed in Table C-11 of *Principles of Chemistry*:

ISOTOPE	RELATIVE MASS	ABUNDANCE
^{204}Pb	203.973020	0.014
^{206}Pb	205.974440	0.241
^{207}Pb	206.975872	0.221
^{208}Pb	207.976627	0.524

A mass spectrum of lead thus will contain four lines, one signal corresponding to each of the four isotopes. The relative intensities will conform to the isotopic distribution at natural abundance:

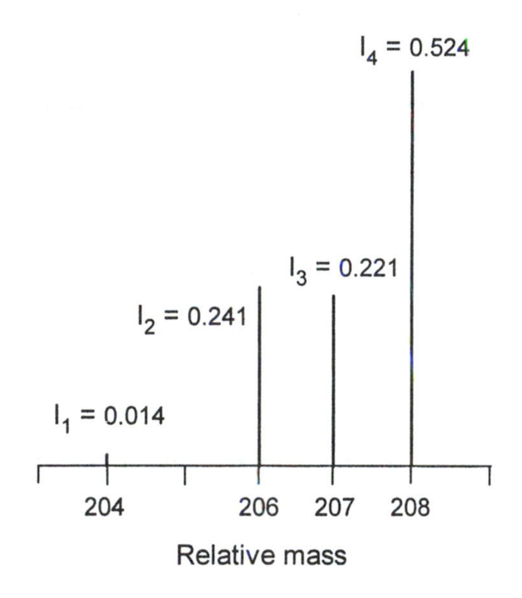

S20-7. The line at a mass/charge ratio of 254, corresponding to the molecular ion $[C_{18}H_{38}]^+$, is unlikely to be the most intense signal in the spectrum. Violently ionized molecules usually fragment into smaller pieces, and only relatively few molecules in the sample remain intact.

S20-8. Structures with a permanent electric dipole moment are microwave-active.

(a) Water is microwave-active. The molecule shows a permanent dipole moment along the HOH bisector:

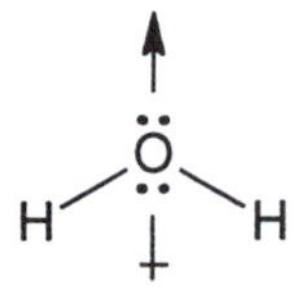

(b) Methane is a symmetric, tetrahedral molecule with no permanent dipole moment:

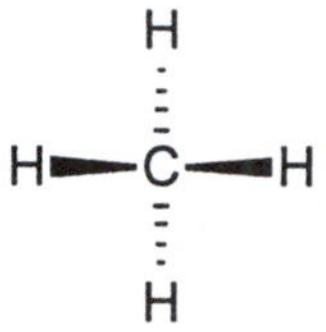

It is not microwave-active.

(c) HCl, a polar diatomic molecule, is microwave-active.

(d) Carbon dioxide, a linear molecule, has no permanent dipole moment and hence is not microwave-active:

$O{=}C{=}O$

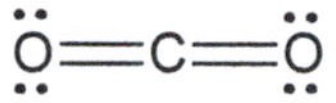

(e) Ammonia is a trigonal pyramidal molecule with a permanent dipole moment along the axis of the pyramid:

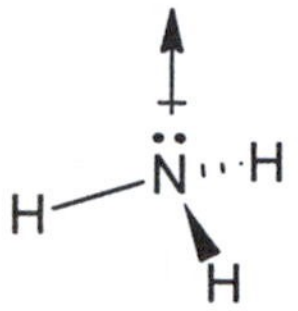

It is microwave-active.

(f) Argon, an atom, possesses spherical symmetry and thus has no permanent dipole moment. It is not microwave-active.

(g) Chloromethane exists as a distorted tetrahedron:

It possesses a permanent dipole moment and consequently is microwave-active.

S20-9. Of the three choices given, only O_2 is paramagnetic in its ground state:

H_2	$(\sigma_{1s})^2$
Li_2	$(\sigma_{2s})^2$
O_2	$(\sigma_{2s})^2(\sigma^*_{2s})^2(\sigma_{2p_z})^2(\pi_{2p})^4(\pi^*_{2p_x})^1(\pi^*_{2p_y})^1$

Molecular oxygen, but not molecular hydrogen or dilithium, is capable of producing a rotational spectrum when it interacts with a magnetic field.

S20-10. Look for an alteration of the charge distribution during the course of a vibration.

(a) The separation between charges varies periodically as the internuclear distances expand and contract. Each $H^{\delta+}$–$O^{\delta-}$ dipole varies in response, and the net molecular dipole moment changes along the H–O–H bisector:

H_2O becomes infrared-active as a result.

(b) During a bending vibration, each bond dipole moment in H_2O undergoes a periodic change in direction. The net vector sum grows and shrinks along the molecular bisector, making this mode of vibration infrared-active as well:

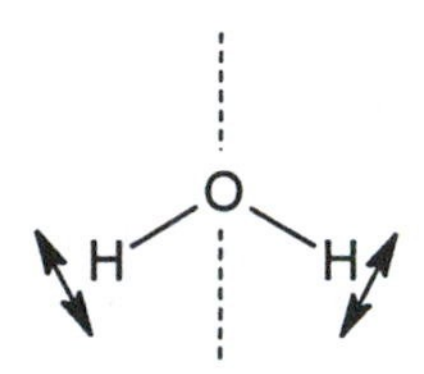

(c) Stretching of the nonpolar H–H bond does not produce a dipole moment. The charge distribution remains symmetric at all times:

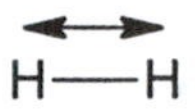

H—H

H_2 is not infrared-active:

(d) HCl, a polar diatomic molecule, develops a changing dipole moment as the bond stretches. The separation between the charges periodically expands and contracts, rendering the molecule infrared-active:

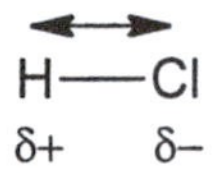

H—Cl
δ+ δ–

(e) A symmetric stretching of the bonds in CO_2 leaves the molecule nonpolar at all times. The two C=O bond dipoles remain diametrically opposed throughout the vibration:

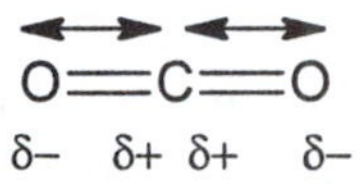

O═C═O
δ– δ+ δ+ δ–

This particular mode is not infrared-active.

(f) A bending vibration in CO_2 causes the bond dipole moments to change direction periodically. Since the OCO bond angle is no longer fixed at 180°, the molecule develops a net dipole moment during the course of the vibration:

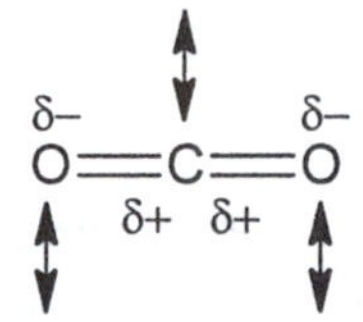

δ– δ–
O═C═O
δ+ δ+

Such modes are infrared-active.

S20-11. Apply the fundamental equations $\lambda\nu = c$ and $E = h\nu$.

(a) A wavelength of 58.43 nm (584.3 Å) falls in the ultraviolet portion of the electromagnetic spectrum.

(b) Frequency is inversely proportional to wavelength:

$$\nu = \frac{c}{\lambda} = \frac{2.998 \times 10^8 \text{ m s}^{-1}}{58.43 \text{ nm} \times \dfrac{1 \text{ m}}{10^9 \text{ nm}}} = 5.131 \times 10^{15} \text{ s}^{-1}$$

(c) Energy, too, is inversely proportional to wavelength:

$$E = h\nu = \frac{hc}{\lambda} = \frac{(6.626 \times 10^{-34}\ \text{J s})(2.998 \times 10^{8}\ \text{m s}^{-1})}{5.843 \times 10^{-8}\ \text{m}} = 3.400 \times 10^{-18}\ \text{J}$$

S20-12. A simple application of ultraviolet photoelectron spectroscopy.

(a) The highest occupied molecular orbital is σ_{2p_z}:

$$\text{N}_2 \qquad (\sigma_{2s})^2(\sigma^{*}_{2s})^2(\pi_{2p})^4(\sigma_{2p_z})^2$$

(b) From the preceding exercise we know how to compute the energy of He(I) radiation:

$$E = h\nu = \frac{hc}{\lambda} = \frac{(6.626 \times 10^{-34}\ \text{J s})(2.998 \times 10^{8}\ \text{m s}^{-1})}{5.843 \times 10^{-8}\ \text{m}} = 3.400 \times 10^{-18}\ \text{J}$$

Next, we compare this value with the stated ionization energy of 1504 kJ mol^{-1}:

$$\frac{1504\ \text{kJ}}{\text{mol}} \times \frac{1\ \text{mol}}{6.022 \times 10^{23}} \times \frac{1000\ \text{J}}{\text{kJ}} = 2.498 \times 10^{-18}\ \text{J}$$

The difference is retained by the ionized electron as kinetic energy:

$$E_\text{k} = (3.400 - 2.498) \times 10^{-18}\ \text{J} = 9.02 \times 10^{-19}\ \text{J}$$

(c) Given the kinetic energy, $E_\text{k} = \frac{1}{2}mv^2$, we solve for the speed v:

$$v = \sqrt{\frac{2E_\text{k}}{m}} = \sqrt{\frac{2(9.02 \times 10^{-19}\ \text{kg m}^2\ \text{s}^{-2})}{9.109 \times 10^{-31}\ \text{kg}}} = 1.41 \times 10^{6}\ \text{m s}^{-1}$$

S20-13. UV photoelectron spectroscopy, continued.

(a) Insert the stated velocity into the defining equation for kinetic energy:

$$E_\text{k} = \frac{1}{2}mv^2 = \frac{1}{2}(9.109 \times 10^{-31}\ \text{kg})(1.262 \times 10^{6}\ \text{m s}^{-1})^2 = 7.254 \times 10^{-19}\ \text{J}$$

Recall that 1 J = 1 kg m^2 s^{-2}.

(b) Subtract the kinetic energy from the photon energy to obtain I, the ionization energy:

$$I = E_{\text{He(I)}} - E_{\text{k}} = \left[\left(3.400 \times 10^{-18}\ \text{J}\right) - \left(7.254 \times 10^{-19}\ \text{J}\right)\right] \times \frac{1\ \text{kJ}}{1000\ \text{J}} \times \frac{6.022 \times 10^{23}}{\text{mol}}$$

$$= 1611\ \text{kJ mol}^{-1}$$

S20-14. Ionization energies of the core electrons in most atoms and molecules usually exceed the energy of an ultraviolet photon. To study a core electron by photoelectron spectroscopy, we typically need to use X rays. See the next exercise.

S20-15. We calculate the wavelength corresponding to the $1s$ ionization energy, as shown below for lithium:

$$E = h\nu = \frac{hc}{\lambda}$$

$$\lambda = \frac{hc}{E} = \frac{\left(6.626 \times 10^{-34}\ \text{J s}\right)\left(2.998 \times 10^{8}\ \text{m s}^{-1}\right)}{\dfrac{4.82 \times 10^{3}\ \text{kJ}}{\text{mol}} \times \dfrac{1\ \text{mol}}{6.022 \times 10^{23}} \times \dfrac{1000\ \text{J}}{\text{kJ}}} = 2.48 \times 10^{-8}\ \text{m}$$

The remaining computations are done in the same way.

	Atom	Ionization Energy (kJ mol^{-1})	Wavelength (m)	Classification
(a)	Li	4.82×10^{3}	2.48×10^{-8}	high-energy ultraviolet (24.8 nm, 248 Å)
(b)	B	1.83×10^{4}	6.54×10^{-9}	X ray (6.54 nm, 65.4 Å)
(c)	F	6.66×10^{4}	1.80×10^{-9}	X ray (1.80 nm, 18.0 Å)

Chapter 21

Worlds Within Worlds—The Nucleus and Beyond

S21-1. Relativistic effects usually scale in proportion to a factor

$$\gamma = \frac{1}{\sqrt{1 - \frac{v^2}{c^2}}}$$

in which c denotes the speed of light and v denotes the velocity of a uniformly moving reference frame. **(a)** Evaluate γ for a particle at rest in a given inertial frame. **(b)** Recalculate γ for the same particle moving with velocity $v = 3 \times 10^6$ m s^{-1}, one-hundredth the speed of light. **(c)** Repeat for $v = 0.1c$, $0.5c$, $0.9c$, $0.99c$, and $0.999c$.

S21-2. The relativistic kinetic energy of a particle with uniform velocity v is given by the following formula:

$$E = \frac{mc^2}{\sqrt{1 - \frac{v^2}{c^2}}} \approx mc^2 + \frac{1}{2}mv^2 + \cdots$$

The *rest mass* of the particle, m, is constant in all inertial reference frames. **(a)** Calculate E for a free electron moving at $0.001c$. **(b)** Calculate the Einsteinian *rest energy* of the particle, mc^2. **(c)** Calculate the nonrelativistic (Newtonian) kinetic energy, $\frac{1}{2}mv^2$. **(d)** How well is the relativistic energy at this speed approximated by the sum of the rest energy and the nonrelativistic kinetic energy?

S21-3. Repeat the preceding exercise for a free electron traveling now at $0.5c$. **(a)** Calculate the relativistic energy. **(b)** Calculate the rest energy. **(c)** Calculate the nonrelativistic kinetic energy. **(d)** Compare the relativistic energy at this speed with the sum of the rest energy and kinetic energy.

S21-4. Once more, this time for a free electron traveling at $0.999c$. **(a)** Calculate the relativistic energy. **(b)** Calculate the rest energy. **(c)** Calculate the nonrelativistic kinetic energy. **(d)** Compare the relativistic energy with the sum of the rest energy and nonrelativistic kinetic energy.

S21-5. **(a)** For which electron—a $1s$ electron in carbon or a $1s$ electron in gold—would relativistic effects be more likely to appear? **(b)** Again: For which electron—a $1s$ electron in gold or a $7s$ electron in plutonium—would relativistic effects be more likely to appear?

S21-6. Suppose that a particle has a nonzero rest mass. What would be its relativistic energy if it moved at the speed of light? Can it do so?

S21-7. The photon has zero rest mass and moves at the speed of light: $v = c$. Yet despite its zero *rest* mass, the photon is still subject to gravitational interactions. A beam of light, for example, will bend in the presence of a strong gravitational field, such as might be produced by a dense star. Are you surprised, given that we customarily think of gravity as a "force" arising from the interaction of two *masses*? Use a very rough, very brief relativistic argument to make plausible the idea that a "massless" particle can be affected by gravity.

S21-8. **(a)** Calculate Δm and ΔE for the nuclear fusion reaction given below:

$$^{1}_{1}\text{H} + {}^{1}_{1}\text{H} \rightarrow {}^{2}_{1}\text{H} + {}^{0}_{1}\text{e} + \nu$$

Use the value 2.014102 u for the atomic mass of deuterium. **(b)** Is the process endothermic or exothermic? Is the transformation favorable from a thermodynamic standpoint?

S21-9. Continue with the fusion reaction described in the preceding exercise, with an eye now toward kinetic rather than thermodynamic benchmarks. **(a)** Start with the formula for the Coulomb energy between charges q_1 and q_2 interacting over a distance r:

$$E = \frac{1}{4\pi\varepsilon_0}\frac{q_1 q_2}{r}$$

Use this equation to calculate the potential energy of two hydrogen nuclei separated by 2×10^{-15} m, the distance at which the strong force begins to take hold. Values of the physical constants are tabulated in Appendix C. **(b)** Think of this repulsive Coulomb potential as a kinetic activation barrier. At approximately what temperature would two hydrogen nuclei have sufficient thermal energy to overcome the barrier? (Remember, from Chapter 10, that $\frac{3}{2}k_B T$ is the average translational kinetic energy of a particle.) **(c)** The answer you have just obtained may well be an overestimate, but ask nonetheless: Is the reaction likely to occur under ordinary terrestrial conditions? **(d)** For comparison, calculate the temperature at which the average translational kinetic energy of a particle would be 50 kJ mol^{-1}—a typical activation barrier for a chemical reaction.

S21-10. The element neon, a noble gas, is unreactive chemically. Does its nobility also extend to nuclear reactions? Explain the difference between a chemical reaction and a nuclear reaction.

S21-11. Different isotopes of the same element exhibit nearly the same chemical behavior. Do two isotopes—say uranium-235 and uranium-238—also behave the same way in nuclear reactions? Why or why not?

S21-12. **(a)** Is a typical nuclear reaction affected by the immediate chemical environment around the nucleus? Will a particular nucleus react differently when incorporated into different molecules? **(b)** Can you think of a possible exception to the answer you just gave?

S21-13. **(a)** Calculate Δm and ΔE for the nuclear transformation given below:

$$^{1}_{0}\text{n} + {}^{235}_{92}\text{U} \rightarrow {}^{142}_{56}\text{Ba} + {}^{91}_{36}\text{Kr} + 3\,{}^{1}_{0}\text{n}$$

The atomic masses of barium-142 and krypton-91 are 141.916360 u and 90.923380 u, respectively. Other relevant data will be found in Appendix C. **(b)** What kind of reaction does the equation describe? **(c)** Is the process endothermic or exothermic? Is it favorable in a thermodynamic sense? **(d)** Is the kinetic activation barrier higher or lower here compared with a fusion reaction? Explain.

S21-14. **(a)** Calculate the energy produced by the annihilation of 1.00×10^{-20} g of electrons and 1.00×10^{-20} g of positrons. **(b)** Calculate the annihilation energy for 1.00×10^{-20} g of protons and 1.00×10^{-20} g of antiprotons.

S21-15. Science fiction or fact: If antimatter could be produced economically and in sufficient quantity, might the process of annihilation offer a practical source of energy? **(a)** Calculate the energy needed to illuminate 10 million 60-watt bulbs for an entire year. Note that the watt (W) is a measure of *power*, energy radiated per unit time: $1 \text{ W} = 1 \text{ J s}^{-1}$. **(b)** What total mass of positrons and electrons (expressed in kilograms) would be sufficient to keep the lights burning?

S21-16. An example of the tremendous energy involved in a typical nuclear fusion process: **(a)** The atomic mass of deuterium is 2.014102 u. Calculate the energy released during the particular reaction shown below, expressing the result in units of kJ mol^{-1}:

$$^{2}_{1}\text{H} + {}^{3}_{1}\text{H} \rightarrow {}^{4}_{2}\text{He} + {}^{1}_{0}\text{n}$$

(b) Suppose that water fills a cubical tank 25.0 meters on a side. If all the energy produced in the making of 1.000 mol ^{4}He were absorbed by the water at 25°C, to what temperature would the liquid rise? Assume that H_2O has a density of 1.00 g cm^{-3} and a heat capacity of 75.3 J mol^{-1} K^{-1} at constant pressure. **(c)** By comparison, what would be the final temperature if (as in a typical chemical reaction) the water absorbed 100 kJ?

S21-17. Don't forget! Strictly speaking, mass is *not* conserved in a chemical reaction. Wherever there is a change in energy, there is a change in mass—but the change in mass brought about by a chemical reaction is so slight as to be practically negligible. **(a)** Calculate ΔE and Δm for a typical chemical transformation, the combustion of one molecule of methane:

$$CH_4(g) + 2O_2(g) \rightarrow CO_2(g) + 2H_2O(g)$$

Hint: Assume that $\Delta E \approx \Delta H$, and then work backward from the equation $\Delta E = (\Delta m)c^2$ to determine Δm. Relevant data may be found in Appendix C. **(b)** Express your value of Δm as a percent change relative to the mass of reactants.

S21-18. Consider the chemical reaction

$$N_2O_4(g) \rightarrow 2NO_2(g)$$

How much mass does the system gain or lose as one molecule of N_2O_4 is converted into two molecules of NO_2? Use the same method as in the preceding example.

S21-19. Is it possible for berkelium-244 to decay into lead-207 through a series of alpha and beta emissions?

S21-20. The difference in energy between the nuclear ground states of $^{27}_{12}\text{Mg}$ and $^{27}_{13}\text{Al}$ is 4.18×10^{-13} J. **(a)** By emission of a β^- particle, a nucleus of magnesium-27 decays first to an excited state of aluminum-27. The excited nucleus then drops to the ground state by emitting a γ photon with a frequency of 2.44×10^{20} Hz. Calculate the energy of the β^- particle and the wavelength of the γ photon, assuming (for simplicity) that no energy is lost through other channels. **(b)** In an alternative route of decay, magnesium-27 emits a β^- particle having an energy of 2.80×10^{-13} J. Estimate the energy and wavelength of the γ photon subsequently emitted, making the same simplifying assumption as before. **(c)** Comment on the wavelengths and energies of the γ radiation.

S21-21. Classify each particle as a lepton, baryon, or meson:

(a) $u\tilde{d}$ **(b)** $\tilde{\nu}$ **(c)** $\tilde{u}\tilde{u}\tilde{d}$ **(d)** $^{1}_{0}\text{n}$ **(e)** $^{0}_{1}\text{e}$

S21-22. Identify the particles in each of the following pairs:

(a) $udd,\ \tilde{u}\tilde{u}\tilde{d}$ **(b)** $^{0}_{-1}\text{e},\ ^{1}_{1}\text{p}$ **(c)** $\nu,\ \tilde{\nu}$ **(d)** $udd,\ \tilde{u}\tilde{d}\tilde{d}$

Which pairs can undergo annihilation?

SOLUTIONS

S21-1. The relativistic scaling factor

$$\gamma = \frac{1}{\sqrt{1-\frac{v^2}{c^2}}}$$

scarcely differs from 1 for speeds small compared with c.

A typical calculation (for $v = 3 \times 10^6$ m s^{-1}) is illustrated below:

$$\gamma = \frac{1}{\sqrt{1-\frac{v^2}{c^2}}} = \frac{1}{\sqrt{1-\frac{(3\times 10^6 \text{ m s}^{-1})^2}{(3\times 10^8 \text{ m s}^{-1})^2}}} = \frac{1}{\sqrt{1-(0.01)^2}} = 1.00005$$

Note that the velocity is conveniently expressed as the ratio $\beta = \frac{v}{c}$.

	β	γ
(a)	0	1
(b)	0.01	1.00005
(c)	0.1	1.005
	0.5	1.155
	0.9	2.29
	0.99	7.09
	0.999	22.4

S21-2. Here we compute the kinetic energy

$$E = \gamma mc^2 = \frac{mc^2}{\sqrt{1-\frac{v^2}{c^2}}} \approx mc^2 + \frac{1}{2}mv^2 + \cdots$$

of a particle moving at a relativistically low speed.

(a) First, calculate γ given that $\frac{v}{c} = 0.001$:

$$\gamma = \frac{1}{\sqrt{1-\frac{v^2}{c^2}}} = \frac{1}{\sqrt{1-(0.001)^2}} = 1.0000005$$

Next, calculate E without making any numerical approximations:

$$\begin{aligned} E = \gamma mc^2 &= (1.0000005)(9.11\times10^{-31}\ \text{kg})(3.00\times10^{8}\ \text{m s}^{-1})^2 \\ &= (1+0.0000005)(8.20\times10^{-14}\ \text{kg m}^2\ \text{s}^{-2}) \\ &= (8.20\times10^{-14}\ \text{J})+(4.10\times10^{-20}\ \text{J}) \end{aligned}$$

(b) The Einsteinian rest energy, mc^2, is obtained by substituting $\gamma = 1$:

$$mc^2 = 8.20\times10^{-14}\ \text{J}$$

(c) Substitute the velocity

$$v = 0.001c = (0.001)(3.00\times10^{8}\ \text{m s}^{-1}) = 3.00\times10^{5}\ \text{m s}^{-1}$$

into the equation for Newtonian kinetic energy:

$$\frac{1}{2}mv^2 = \frac{1}{2}(9.11\times10^{-31}\ \text{kg})(3.00\times10^{5}\ \text{m s}^{-1})^2 = 4.10\times10^{-20}\ \text{J}$$

(d) Compare the numbers calculated in (b) and (c) with the number calculated in (a). The approximation

$$E \approx mc^2 + \frac{1}{2}mv^2 = (8.20\times10^{-14}\ \text{J})+(4.10\times10^{-20}\ \text{J})$$

holds to better than 21 decimal places.

S21-3. Use the same method as in the preceding exercise, this time for $\frac{v}{c} = 0.5$:

$$v = 0.5c = 1.50\times10^{8}\ \text{m s}^{-1}$$

(a) Here the factor γ differs noticeably from its low-velocity limit of 1,

$$\gamma = \frac{1}{\sqrt{1-\frac{v^2}{c^2}}} = \frac{1}{\sqrt{1-(0.5)^2}} = 1.1547$$

and the energy scales in direct proportion:

$$E = \gamma mc^2 = (1.1547)(9.11 \times 10^{-31} \text{ kg})(3.00 \times 10^8 \text{ m s}^{-1})^2 = 9.47 \times 10^{-14} \text{ J}$$

(b) The rest energy, an intrinsic property of the particle, is the same regardless of the velocity v:

$$mc^2 = (9.11 \times 10^{-31} \text{ kg})(3.00 \times 10^8 \text{ m s}^{-1})^2 = 8.20 \times 10^{-14} \text{ J}$$

(c) The Newtonian kinetic energy is given, as usual, by $\frac{1}{2}mv^2$:

$$\frac{1}{2}mv^2 = \frac{1}{2}(9.11 \times 10^{-31} \text{ kg})(1.50 \times 10^8 \text{ m s}^{-1})^2 = 1.02 \times 10^{-14} \text{ J}$$

(d) Terms of order higher than v^2 contribute just over 2.5% of the total energy:

$$mc^2 + \frac{1}{2}mv^2 = (8.20 \times 10^{-14} \text{ J}) + (1.02 \times 10^{-14} \text{ J}) = 9.22 \times 10^{-14} \text{ J}$$

$$\frac{mc^2 + \frac{1}{2}mv^2}{E} = \frac{9.22 \times 10^{-14} \text{ J}}{9.47 \times 10^{-14} \text{ J}} = 0.974$$

S21-4. Take the same approach as in the preceding two exercises. With the velocity now approaching the speed of light ($v = 0.999c$), the scaling factor γ begins to rise steeply:

$$\gamma = \frac{1}{\sqrt{1-\frac{v^2}{c^2}}} = \frac{1}{\sqrt{1-(0.999)^2}} = 22.366$$

Results are summarized below:

(a) $E = \gamma mc^2 = 1.83 \times 10^{-12}$ J

(b) $mc^2 = 8.20 \times 10^{-14}$ J

(c) $\frac{1}{2}mv^2 = 4.09 \times 10^{-14}$ J

(d) $$\frac{mc^2 + \frac{1}{2}mv^2}{E} = 0.067$$

S21-5. Relativistic effects become significant only at high kinetic energies and high speeds, such as experienced by an electron near a bare nucleus. Typically the speed of an inner-shell electron increases as the nuclear charge increases.

(a) A $1s$ electron in gold, subject to a larger nuclear charge, is more likely to behave relativistically than a $1s$ electron in carbon.

(b) A $1s$ electron in gold—close to the nucleus—is more likely to behave relativistically than a shielded $7s$ valence electron in plutonium.

S21-6. A particle with nonzero rest mass cannot achieve a velocity equal to c. If it did, its relativistic energy would be infinite:

$$E = \frac{mc^2}{\sqrt{1 - \frac{v^2}{c^2}}} \rightarrow \frac{mc^2}{0} \qquad \text{as} \qquad v \rightarrow c$$

S21-7. According to the special theory of relativity, the mass–energy relationship

$$E = mc^2 \qquad (v = 0)$$

shows that inertial mass is a form of energy, and vice versa. A particle with zero rest mass still has a nonzero energy pc, and this finite *mass–energy* is subject to a gravitational field.

According to the general theory of relativity (Einstein's picture of gravitation), space-time is curved in the vicinity of a massive object. A small test particle, rather than responding to the "force" of a gravitational field, simply follows the local curvature: It moves inertially (force-free) along a bent path, traveling the shortest possible distance in the curved four-dimensional space. Even a photon, a particle with zero rest mass, deviates from a straight-line path in the presence of gravitational masses.

S21-8. In this exercise and the next, we compare the favorable thermodynamics of a typical fusion reaction with its distinctly unfavorable kinetics. A huge amount of energy stands waiting to be released (see below), but the activation barrier is formidable (see Exercise S21-9).

Relevant particle and atomic masses are listed in Tables C-5 and C-11 of the text, as well as on page 768. The mass of a deuterium atom is given specially as 2.014102 u.

(a) Two hydrogen nuclei fuse into a single deuterium nucleus, releasing a positron and a neutrino:

$$ {}_{1}^{1}\text{H} + {}_{1}^{1}\text{H} \rightarrow {}_{1}^{2}\text{H} + {}_{1}^{0}\text{e} + \nu $$

The mass of a hydrogen nucleus is equal to the mass of a proton, $m_{\text{p}} = 1.00727647$ u. We then take the mass of a positron as equal to the mass of an electron (m_{e}), and we assume further that the neutrino has zero rest mass. Subtracting m_{e} from the atomic mass of deuterium (which ordinarily includes both nucleus and electron), we obtain the overall decrease in mass:

$$ \begin{aligned} \Delta m &= m({}_{1}^{2}\text{H}) + m({}_{1}^{0}\text{e}) - 2m({}_{1}^{1}\text{H}) \\ &= (2.014102\ \text{u} - m_{\text{e}}) + m_{\text{e}} - 2m_{\text{p}} \\ &= 2.014102\ \text{u} - 2(1.00727647\ \text{u}) \\ &= -0.000451\ \text{u} \end{aligned} $$

The change in energy follows directly from Einstein's mass–energy equation:

$$ \Delta E = (\Delta m)c^2 = \left(-0.000451\ \text{u} \times \frac{1.6605402 \times 10^{-27}\ \text{kg}}{\text{u}}\right)\left(2.99792458 \times 10^{8}\ \text{m s}^{-1}\right)^2 $$

$$ \times \frac{1\ \text{J}}{\text{kg m}^2\ \text{s}^{-2}} \times \frac{6.0221367 \times 10^{23}}{\text{mol}} \times \frac{1\ \text{kJ}}{1000\ \text{J}} = -4.053 \times 10^{7}\ \text{kJ mol}^{-1} $$

(b) The transformation is highly exothermic and thermodynamically favorable.

S21-9. A *rough* estimate of the barrier to fusion, based on an electrostatic model.

(a) Substitute the specified values of charge and distance into the Coulomb formula:

$$ E = \frac{1}{4\pi\varepsilon_0}\frac{q_1 q_2}{r} = \frac{1}{4\pi\left(8.854 \times 10^{-12}\ \text{C}^2\ \text{N}^{-1}\ \text{m}^{-2}\right)} \times \frac{\left(1.602 \times 10^{-19}\ \text{C}\right)^2}{2 \times 10^{-15}\ \text{m}} = 1.15 \times 10^{-13}\ \text{N m} $$

$$ = 1.15 \times 10^{-13}\ \text{J} \quad (\approx 7 \times 10^{7}\ \text{kJ mol}^{-1}) $$

(b) Insert the number just obtained into the equation for thermal energy,

$$E = \frac{3}{2} k_B T$$

and solve for the absolute temperature T—over 5 billion kelvins:

$$T = \frac{2E}{3k_B} = \frac{2\left(1.15 \times 10^{-13}\ \text{J}\right)}{3\left(1.38066 \times 10^{-23}\ \text{J K}^{-1}\right)} = 5.6 \times 10^9\ \text{K}$$

(c) The fusion reaction is extremely unlikely to occur under normal terrestrial conditions. Its energy of activation far exceeds the thermal energy typically available.

(d) Repeat the calculation made in (b), this time for an energy of only 50 kJ mol^{-1}:

$$T = \frac{2E}{3R} = \frac{2\left(50\ \text{kJ mol}^{-1}\right)}{3\left(8.3145 \times 10^{-3}\ \text{kJ mol}^{-1}\ \text{K}^{-1}\right)} \approx 4000\ \text{K}$$

Recall that $R = N_0 k_B$, where N_0 is Avogadro's number.

S21-10. A chemical reaction typically involves the valence electrons of atoms and molecules, whereas a nuclear reaction involves the nucleons inside a nucleus. The energy exchanged in a nuclear reaction is vastly greater than in a chemical reaction.

The chemical "nobility" of a neon atom—manifested in its nonreactive valence shell—therefore has no bearing on the internal transformation of the atom's nucleus.

S21-11. As in Exercise S21-10, no: Uranium-235 and uranium-238 behave entirely differently in nuclear reactions. Their nucleonic structure and binding energies are not the same. The similarity of their electron distributions is irrelevant.

S21-12. Nuclear reactions versus atomic and molecular reactions, continued.

(a) Again, most nuclear reactions are unaffected by the chemical (electronic) environment around the nucleus.

(b) An exception to the general rule is provided by the process of *electron capture*, which directly involves the "chemical" electrons outside the nucleus.

S21-13. Uranium-235 absorbs a neutron and undergoes fission:

$${}^{1}_{0}\text{n} + {}^{235}_{92}\text{U} \rightarrow {}^{142}_{56}\text{Ba} + {}^{91}_{36}\text{Kr} + 3\,{}^{1}_{0}\text{n}$$

The reaction under consideration represents just one of many possible pathways.

Relevant particle and atomic masses are listed in Tables C-5 and C-12 of *Principles of Chemistry*, as well as on page 768.

(a) Calculate, first, the change in mass:

$$\begin{aligned}\Delta m &= m\left({}^{142}_{56}\mathrm{Ba}\right) + m\left({}^{91}_{36}\mathrm{Kr}\right) + 3m\left({}^{1}_{0}\mathrm{n}\right) - m\left({}^{1}_{0}\mathrm{n}\right) - m\left({}^{235}_{92}\mathrm{U}\right)\\ &= \left(141.916360\ \mathrm{u} - 56m_\mathrm{e}\right) + \left(90.923380\ \mathrm{u} - 36m_\mathrm{e}\right) + 2m_\mathrm{n} - \left(235.043924\ \mathrm{u} - 92m_\mathrm{e}\right)\\ &= 141.916360\ \mathrm{u} + 90.923380\ \mathrm{u} + 2\left(1.00866490\ \mathrm{u}\right) - 235.043924\ \mathrm{u}\\ &= -0.186854\ \mathrm{u}\end{aligned}$$

And then the change in energy:

$$\Delta E = (\Delta m)c^2 = \left(-0.186854\ \mathrm{u} \times \frac{1.6605402 \times 10^{-27}\ \mathrm{kg}}{\mathrm{u}}\right)\left(2.99792458 \times 10^{8}\ \mathrm{m\,s^{-1}}\right)^2$$

$$= -2.789 \times 10^{-11}\ \mathrm{J} \quad \left(-1.679 \times 10^{10}\ \mathrm{kJ\ mol^{-1}}\right)$$

(b) The transformation is classified as nuclear fission: A heavy nucleus (^{235}U) decomposes into two lighter nuclei (^{142}Ba and ^{91}Kr), releasing three neutrons in the process.

(c) The reaction is highly exothermic and thermodynamically favorable.

(d) The kinetic activation barrier is lower in comparison to a fusion reaction, because the uncharged neutron can approach the positively charged uranium-235 nucleus without suffering electrostatic repulsion.

S21-14. The annihilation of each pair converts two particle masses into an equivalent amount of energy.

(a) Use Einstein's equation $\Delta E = (\Delta m)c^2$ to obtain the annihilation energy for a total mass of $2 \times (1.00 \times 10^{-20}$ g):

$$\Delta E = (\Delta m)c^2 = \left(-2.00 \times 10^{-20}\ \mathrm{g} \times \frac{1\ \mathrm{kg}}{1000\ \mathrm{g}}\right)\left(2.998 \times 10^{8}\ \mathrm{m\ s^{-1}}\right)^2 = -1.80 \times 10^{-6}\ \mathrm{J}$$

Recall that $1\ \mathrm{J} = 1\ \mathrm{kg\ m^2\ s^{-2}}$.

(b) The total change in mass—and hence the total change in energy—is the same as in (a): $\Delta E = -1.80 \times 10^{-6}$ J.

S21-15. Mass and energy, continued.

(a) Energy is the product of power and time:

$$\overset{\text{POWER}}{E = \left(\frac{60\ \text{J s}^{-1}}{\text{bulb}} \times 10^{7}\ \text{bulbs}\right)}\overset{\text{TIME}}{\left(1\ \text{y} \times \frac{365.25\ \text{d}}{\text{y}} \times \frac{86{,}400\ \text{s}}{\text{d}}\right)} = 1.9 \times 10^{16}\ \text{J}$$

(b) Use the mass–energy equation to determine the remarkably small amount of mass required—less than half a pound:

$$\Delta m = \frac{\Delta E}{c^2} = \frac{1.9 \times 10^{16}\ \text{kg m}^2\ \text{s}^{-2}}{\left(3.00 \times 10^{8}\ \text{m s}^{-1}\right)^2} = 0.21\ \text{kg}$$

S21-16. Deuterium and tritium fuse together to form helium-4, generating a neutron as a by-product:

$$^{2}_{1}\text{H} + {}^{3}_{1}\text{H} \rightarrow {}^{4}_{2}\text{He} + {}^{1}_{0}\text{n}$$

The atomic mass of deuterium is stated in the problem as 2.014102 u. See Tables C-5, C-11, and C-12 in the text, as well as page 768, for the other relevant masses.

(a) The combined mass of the products is less than the combined mass of the reactants:

$$\begin{aligned}
\Delta m &= m\left({}^{4}_{2}\text{He}\right) + m\left({}^{1}_{0}\text{n}\right) - m\left({}^{2}_{1}\text{H}\right) - m\left({}^{3}_{1}\text{H}\right) \\
&= \left(4.002603\ \text{u} - 2m_{\text{e}}\right) + \left(1.0086649\ \text{u}\right) - \left(2.014102\ \text{u} - m_{\text{e}}\right) - \left(3.01605\ \text{u} - m_{\text{e}}\right) \\
&= 4.002603\ \text{u} + 1.0086649\ \text{u} - 2.014102\ \text{u} - 3.01605\ \text{u} \\
&= -0.018884\ \text{u}
\end{aligned}$$

The difference in mass is manifested as energy:

$$\Delta E = (\Delta m)c^2 = \left(-0.018884\ \text{u} \times \frac{1.6605402 \times 10^{-27}\ \text{kg}}{\text{u}}\right)\left(2.99792458 \times 10^{8}\ \text{m s}^{-1}\right)^2$$

$$\times \frac{1\ \text{J}}{\text{kg m}^2\ \text{s}^{-2}} \times \frac{1\ \text{kJ}}{1000\ \text{J}} \times \frac{6.0221367 \times 10^{23}}{\text{mol}} = -1.697 \times 10^{9}\ \text{kJ mol}^{-1}$$

(b) Use the equation $q_P = nc_P\,\Delta T$, where n denotes the molar amount of H_2O, c_P denotes the heat capacity at constant pressure, and ΔT denotes the change in temperature:

$$q_P = \frac{1.697\times10^9\ \text{kJ}}{\text{mol}} \times \frac{1000\ \text{J}}{\text{kJ}} \times 1.000\ \text{mol} = 1.697\times10^{12}\ \text{J}$$

$$n = \left(25.0\ \text{m} \times \frac{100\ \text{cm}}{\text{m}}\right)^3 \times \frac{1.00\ \text{g}}{\text{cm}^3} \times \frac{1\ \text{mol}}{18.015\ \text{g}} = 8.673\times10^8\ \text{mol}$$

$$c_P = 75.3\ \text{J mol}^{-1}\ \text{K}^{-1}$$

Heat is transferred from the exothermic nuclear reaction to the reservoir of water, thereby raising its temperature. Solving for ΔT, we find that the final temperature is 51.0°C—a huge increase, in view of the large quantity of water:

$$\Delta T = \frac{q_P}{nc_P} = \frac{1.697\times10^{12}\ \text{J}}{(8.673\times10^8\ \text{mol})(75.3\ \text{J mol}^{-1}\ \text{K}^{-1})} = 26.0\ \text{K}$$

$$T_2 = T_1 + \Delta T = 25.0^\circ\text{C} + 26.0^\circ\text{C} = 51.0^\circ\text{C}$$

(c) If $q_P = 10^5$ J in the equation above, then $\Delta T = 1.53\times10^{-6}$ K. The change in temperature will be imperceptible, leaving the system at ≈25.0°C.

S21-17. We apply the mass–energy equation to an ordinary combustion reaction:

$$CH_4(g) + 2O_2(g) \rightarrow CO_2(g) + 2H_2O(g)$$

(a) Making the approximation $\Delta E \approx \Delta H$, we first calculate the enthalpy of reaction:

$$\Delta H^\circ = \Delta H_f^\circ[CO_2(g)] + 2\,\Delta H_f^\circ[H_2O(g)] - \Delta H_f^\circ[CH_4(g)] - 2\,\Delta H_f^\circ[O_2(g)]$$

$$= \left(-\frac{393.5\ \text{kJ}}{\text{mol}} \times 1\ \text{mol}\right) + \left(-\frac{241.8\ \text{kJ}}{\text{mol}} \times 2\ \text{mol}\right) - \left(-\frac{74.8\ \text{kJ}}{\text{mol}} \times 1\ \text{mol}\right) - \left(\frac{0\ \text{kJ}}{\text{mol}} \times 2\ \text{mol}\right)$$

$$= -\frac{802.3\ \text{kJ}}{\text{mol CH}_4} \times \frac{1\ \text{mol CH}_4}{6.022\times10^{23}} \times \frac{1000\ \text{J}}{\text{kJ}} = -1.332\times10^{-18}\ \text{J}\quad(\text{per molecule CH}_4)$$

The equivalent mass follows from the Einstein equation:

$$\Delta m = \frac{\Delta E}{c^2} = -\frac{1.332\times10^{-18}\ \text{J}}{(2.998\times10^8\ \text{m s}^{-1})^2} \times \frac{1\ \text{kg m}^2\ \text{s}^{-2}}{\text{J}} = -1.482\times10^{-35}\ \text{kg}$$

(b) The decrease in mass is negligible, barely 0.00000001% of the original mass of reactants:

$$\frac{\Delta m}{m(CH_4) + 2m(O_2)} \equiv -\frac{1.482 \times 10^{-35} \text{ kg} \times \dfrac{1 \text{ u}}{1.66054 \times 10^{-27} \text{ kg}}}{16.043 \text{ u} + 2(31.9988 \text{ u})} \times 100\% = -(1.115 \times 10^{-8})\%$$

S21-18. Since the reaction

$$N_2O_4(g) \rightarrow 2NO_2(g)$$

is endothermic,

$$\Delta H^\circ = 2\,\Delta H_f^\circ[NO_2(g)] - \Delta H_f^\circ[N_2O_4(g)] = \left(\frac{33.2 \text{ kJ}}{\text{mol}} \times 2 \text{ mol}\right) - \left(\frac{9.2 \text{ kJ}}{\text{mol}} \times 1 \text{ mol}\right)$$

$$= 57.2 \text{ kJ}$$

the system *gains* mass:

$$\Delta m = \frac{\Delta E}{c^2} \approx \frac{\Delta H^\circ}{c^2} = \frac{\dfrac{57{,}200 \text{ J}}{\text{mol } N_2O_4} \times \dfrac{1 \text{ mol } N_2O_4}{6.022 \times 10^{23}}}{(2.998 \times 10^8 \text{ m s}^{-1})^2} \times \frac{1 \text{ kg m}^2 \text{ s}^{-2}}{\text{J}} = 1.06 \times 10^{-36} \text{ kg}$$

We make the approximation that $\Delta E \approx \Delta H^\circ$.

S21-19. The question is whether the transformation

$$^{244}_{97}\text{Bk} \rightarrow {}^{207}_{82}\text{Pb}$$

can be realized by emission of alpha and beta particles. It cannot: The difference in mass number,

$$244 - 207 = 37,$$

is not a multiple of 4 (the mass of an alpha particle). Emission of beta particles has no effect on the mass number.

S21-20. We apply the familiar relationships connecting electromagnetic wavelength, frequency, and energy:

$$\lambda\nu = c$$

$$E = h\nu = \frac{hc}{\lambda}$$

(a) See the diagram below:

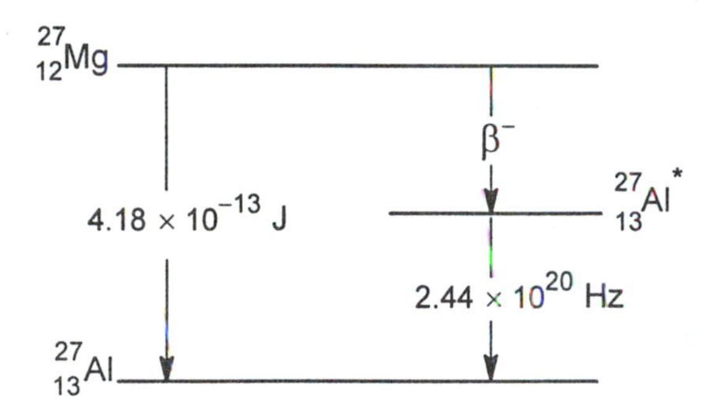

From the frequency of the γ photon (2.44×10^{20} Hz) we determine the associated wavelength and energy:

$$\lambda_\gamma = \frac{c}{\nu_\gamma} = \frac{2.998 \times 10^8 \text{ m s}^{-1}}{2.44 \times 10^{20} \text{ s}^{-1}} = 1.23 \times 10^{-12} \text{ m}$$

$$E_\gamma = h\nu_\gamma = \left(6.626 \times 10^{-34} \text{ J s}\right)\left(2.44 \times 10^{20} \text{ s}^{-1}\right) = 1.62 \times 10^{-13} \text{ J}$$

The energy of the β^- particle is then established as follows:

$$E_{\beta^-} = 4.18 \times 10^{-13} \text{ J} - E_\gamma = 2.56 \times 10^{-13} \text{ J}$$

(b) Given an energy of 2.80×10^{-13} J for the β^- particle, we estimate the energy and wavelength of the accompanying γ photon:

$$E_\gamma = 4.18 \times 10^{-13} \text{ J} - E_{\beta^-} = 4.18 \times 10^{-13} \text{ J} - 2.80 \times 10^{-13} \text{ J} = 1.38 \times 10^{-13} \text{ J}$$

$$\lambda_\gamma = \frac{hc}{E_\gamma} = \frac{\left(6.626 \times 10^{-34} \text{ J s}\right)\left(2.998 \times 10^8 \text{ m s}^{-1}\right)}{1.38 \times 10^{-13} \text{ J}} = 1.44 \times 10^{-12} \text{ m}$$

(c) Gamma rays, with exceedingly small wavelengths, are the most energetic form of electromagnetic radiation. The photons considered in this exercise, for example, have wavelengths between 100 and 1000 times shorter than those of typical X rays (which might fall in the range 10^{-9} m < λ < 10^{-10} m). The γ energies are larger by the same factors.

S21-21. Neutrons and protons, both baryons, are built from three quarks, whereas mesons are composed of a quark and an antiquark. Leptons, such as electrons and neutrinos, have no internal quark structure.

(a) $u\tilde{d}$ meson (π^+)

(b) $\tilde{\nu}$ antilepton (antineutrino)

(c) $\tilde{u}\tilde{u}\tilde{d}$ antibaryon (antiproton)

(d) ${}^{1}_{0}\text{n}$ baryon (neutron, *udd*)

(e) ${}^{0}_{1}\text{e}$ antilepton (positron)

S21-22. Annihilation will occur only if each elementary particle (quark or lepton) is matched by a corresponding antiparticle.

(a) A neutron and an antiproton cannot annihilate each other:

udd	neutron
$\tilde{u}\tilde{u}\tilde{d}$	antiproton

(b) An electron (${}^{0}_{-1}\text{e}$, a lepton) and a proton (${}^{1}_{1}\text{p} = uud$, a baryon) are not eligible for mutual annihilation. The particles are of different classes.

(c) A neutrino (ν) and antineutrino ($\tilde{\nu}$)—a matched lepton–antilepton pair—can undergo annihilation.

(d) A baryon–antibaryon pair, the neutron and antineutron will annihilate each other upon meeting:

udd	neutron
$\tilde{u}\tilde{d}\tilde{d}$	antineutron